THE END

ALSO BY MARQ DE VILLIERS

White Tribe Dreaming

Down the Volga

The Heartbreak Grape

Blood Traitors (with Sheila Hirtle)

Into Africa (with Sheila Hirtle)

Water

Sahara (with Sheila Hirtle)

Sable Island (with Sheila Hirtle)

Windswept

Witch in the Wind

Timbuktu (with Sheila Hirtle)

THE END

Natural Disasters, Manmade
Catastrophes, and the
Future of Human Survival

MARQ DE VILLIERS

THOMAS DUNNE BOOKS

ST. MARTIN'S PRESS ≋ NEW YORK

THOMAS DUNNE BOOKS.
An imprint of St. Martin's Press.

Quotations from Simon Winchester's *A Crack in the Edge of the World* by permission of HarperCollins Publishers. Copyright © 2005 Simon Winchester.

Passages from Richard Fortey's *The Earth: An Intimate History*, reprinted by permission of HarperCollins Publishers. Copyright © 2005 by Richard Fortey.

Quotations from *The Revenge of Gaia* by James Lovelock reprinted by permission of Basic Books, a member of the Perseus Books Group. Copyright © 2006 by James Lovelock.

Passages from Ernest Zebrowski's *Perils of a Restless Planet* reprinted with the permission of Cambridge University Press. Copyright © 1997 by Ernest Zebrowski.

www.thomasdunnebooks.com
www. stmartins.com

Library of Congress Cataloging-in-Publication Data

De Villiers, Marq.
    [Dangerous world]
    The end : natural disasters, manmade catastrophes, and the future of human survival / Marq de Villiers.—1st U.S. ed.
        p. cm.
    "First published in Canada as Dangerous world by Viking Canada"—T.p. verso.
    Includes bibliographical references and index.
    ISBN-13: 978-0-312-36569-1
    ISBN-10: 0-312-36569-1
  1. Natural disasters.   2. Disasters.   3. Human ecology.   4. Emergency management.   I. Title.
   GB5014.D48   2008
   363.34—dc22

                                              2008039096

First published in Canada by Viking Canada, an imprint of the Penguin Group.

First U.S. Edition: December 2008

10 9 8 7 6 5 4 3 2 1

Contents

Part Three: Peril by Peril

Part Four: What Is to Be Done?

PART ONE

So What's the Problem?

ONE

Doomsday As a State of Mind

All kinds of terrible things could happen, and the universe of terrible things is so large that some of them probably will.
—Stephen Pacala, Princeton University ecologist

Well there it is, in a nutshell: doomsday as a state of mind. This is not to say that Pacala was wrong, exactly—he was in any case talking about climate change and its effects on the biosphere, and not about the End of Days—but the style of thinking perfectly matches the anxious modern mood. Consider how accustomed (though not inured) we have all become to the vocabulary of catastrophe. Tsunami, earthquake, volcano, hurricane, pandemic, cosmic and UV radiation, comets in collision—these things used to worry us, from time to time, and certainly they used to panic those directly affected, but they weren't really part of the background noise of our culture. Now they are. The Christmas tsunami that swept through the Indian Ocean in 2004 washed up directly into our living rooms and into our forebrains. We have all seen the hapless victims picking through the rubble after an earthquake in Turkey, Greece, Pakistan, Japan, Peru; the fallen Californian expressway crushing the cars below

3

is replayed every time the Earth shakes somewhere; San Francisco (the "crack in the edge of the world," Simon Winchester called it), Krakatoa, Mount St. Helens, hurricanes Katrina and Rita and Wilma and Dean, a deadly tornado ripping apart a small town in Kansas, the suffocation of all the people in a clutch of villages in Cameroon—we all know all these things, for we've seen them and heard them and watched the wailing and lamentation that inevitably follows, an intimate shorthand of calamity. As the number of intense hurricanes began to increase, as the earthquake-caused tsunami flattened coastal towns all around the Indian Ocean, as a massive earthquake toppled cities on the Roof of the World in Pakistan, as the weather turned more and more bizarre and flooding and droughts and heat waves increased, as the dire warnings kept coming of Avian flu, SARS, Ebola, and assorted pandemics, so the global anxiety levels continued to escalate. Consider how the jargon of meteorology has already become common currency—frontal systems, Category Four storm, Saffir-Simpson scale, isobars, extra-tropical cyclone, and the rest— who knew these things before? Global warming with its array of hitherto obscure gases—methane, chlorofluorocarbons, ozone—has seeped out of the scientific journals to become a basic part of popular discourse. Schoolchildren can now talk knowledgeably about antiretrovirals, bacterial infections, the collapse of immune systems. . . . A survey even showed, bizarrely, that more people were aware of the wandering asteroid that wiped out the dinosaurs than actually believed the dinosaurs had their place in the evolution of planetary life in the first place.

Calamity has become not just a state of mind but a majority state of mind. "The world can now expect three to five major disasters a year that will each kill more than 50,000 people"—when this piece of advice was published in 2005, as an analysis of natural trends by the insurance giant Munich Re, hardly anyone turned a hair. It seemed merely obvious—during the next twelve months, you are almost certainly going to be reading about one or two, perhaps three, natural calamities that each kills many thousands of people. The chances of your being one of the victims are still small, but the chances of there being many victims approaches 100 percent. It has become *normal*.

The *Bulletin of Atomic Scientists* still keeps its doomsday clock, a dismal relic of the dismal Cold War and the prospect of nuclear winter, but when it solemnly advanced the clock to five minutes to midnight, a result, it said, of a worsening climate and increasing nuclear proliferation, most media outlets gave it a paragraph or two, no more. Ho-hum, another disaster forecast.

Perhaps it was always thus, but not in all cultures and not to the same degree. I once asked a member of the Dogon tribe of Mali, of all African tribes perhaps the one with the most complex cosmology interwoven into its religious beliefs, about this notion of the end of the world. I wanted to know how he thought the end would come; he was the grandson of the tribe's most famous seer and would no doubt have something interesting to say. But he merely looked puzzled. Why should it end? People end, yes, so do animals, even the tricky jackal. Good times end, and so do droughts. But the world just is.

Worlds populated by families or tribes of gods typically don't end. The Haida of the American northwest, whose complex myths have been shown to have tracked, and then predicted, natural calamities like earthquakes and tsunamis (as we shall see) nevertheless were unable to imagine everything actually ending. The gods might make wars on each other, and throw each other down, and in their skirmishings people are battered and bruised, but the world persists. The ancient Greeks invented perhaps the most quarrelsome family of gods in world history, gods who could and did bring ruin to people in their battles (Vulcan's forge was Vesuvius, rumbling the world as he made mighty weapons for the endless wars waged by his brothers and sisters), but the world itself was untouched. Self-evidently, cosmologies that included resurrection and eternal self-improvement (Hinduism, Buddhism) could make no room for the End, though they too were not shy of prophecy. For example, after the Indian Ocean tsunami of 2004, a religious leader named Mata Amritanandamayi, usually just known as Amma, attributed the increasing number of natural calamities to a radical decline in *dharma*, or righteousness; her advice was to instruct devotees to leave everything to the will of God and face all difficulties with courage, which seems helpful.

It was left, then, to the mighty monotheisms to codify and attribute apocalypse: "And God said unto Noah, The end of all flesh is come before me; for the Earth is filled with violence through them and behold, I will destroy them with the Earth." Well, he didn't. But he did have a crack at it.

Perhaps this attention to apocalypse is not so surprising. These theisms were born in the Levant and Mesopotamia (what we in our Eurocentric way call the Middle East), a crucible of planetary unrest: the earthquakes, floods, pestilences, and droughts that so frequently occurred there were preserved for millennia in folk memory. If the Jehovah of the Jews was a cruel and vindictive lord, perhaps he was a reflection of the landscape into which he was imagined. It is hard not to see something real in the gloomy recountings of the book of Revelation, the most apocalyptic, not to say paranoid, of all biblical texts:

> And the third angel sounded, and there fell a great star from heaven, burning as if it were a lamp, and it fell upon the third part of the rivers and upon the fountain of the waters; and the name of the star is called wormwood, and the third part of the waters became wormwood, and many men died of the waters, because they were made bitter; and the fourth angel sounded, and the third part of the sun was smitten, and the third part of the moon, and the third part of the stars; so as the third part of them was darkened, and the day shone not for a third part of it, and the night likewise. . . .

Something very like these events did happen, from time to time—recorded not only in the long memories of human scribes but also in the very rocks of the Earth. The calamities they describe would have been profound enough to burn into long tribal memory. Just as the Haida remember the wars between Eagle and Wind, the Old Testament authors remembered the fury of Jehovah: "And lo, there was a great earthquake, and the sun became black as sackcloth of hair, and the moon became as blood; and the stars of heaven fell unto the Earth, even as a fig tree casteth her untimely figs when she is shaken by a mighty wind. . . ." The fig tree is a nice homely touch, giving away this story's Mediterranean origins; maybe this was a kind of reportage, after all.

In the Arabic world, where rebellion against a duly constituted ruler is both common and anathema, prolonged strife, or *fitnah*, is easily mistaken for the imminence of the Day of Judgment, a dire notion Islam shares with the less worldly among the Christians:

When the sun shall be darkened
When the stars shall be thrown down
When the mountains shall be set moving
When the pregnant camels shall be neglected
When the savage beasts shall be mustered
When the seas shall be set boiling
When the souls shall be coupled
When the buried infant shall be asked for what sin she was slain
When the scrolls shall be unrolled
When heaven shall be stripped off
When hell shall be set blazing
When Paradise shall be brought nigh
Then shall a soul know what it has produced.[1]

Christian documents are punctuated by the conflation of disaster with divine wrath. Saint Paul's writing are full of such fear, though he nevertheless rather looked forward to the End of Days, confident it would come in his own time and that he himself would be among the elect of heaven.

Such writings persist into modern times, especially among thinkers less eminent by far than the great disciple. In fact, the most entertaining doomsday speculations come from pseudo-swamis and addle-brained reverends across a dozen cultures—no one is really immune to this stuff. Even in technologically sophisticated America, as Sam Harris points out in his *Letter to a Christian Nation*, "about half our neighbors believe the entire cosmos was created six thousand years ago. This is, incidentally, about a thousand years after the Sumerians invented glue. . . ."[2]

It should be no surprise that the wilder shores of the Internet are rife with conspiracy theories, doomsday scenarios, and the many ways in which scientists are trying to kill us all, or, rather more simply, with speculations that the world is going to end without any help from science—or from Satan, for that matter. There is no point cataloging them, for they are all similar enough in structure. It is interesting, though, that almost all of them, in these days, are underpinned by a "scientific" reading of some ancient or esoteric event.

On the fringes of modern Christianity are those who firmly believe in the Rapture,[3] the gathering up to Jesus of the faithful at the Second Coming, which of course is imminent (as it has been since Saint Paul's day, although the term was actually introduced in the middle of the second century). Most mainstream denominations ignore the whole Rapture thing as a theological embarrassment; it appears in most concordances and biblical commentaries only as a curiosity, usually under a heading such as "eschatological events." The fact that the fictional series called *Left Behind*, by Tim LaHaye and Jerry B. Jenkins, has sold gazillions of copies without the benefit of bookstores is usually cited as evidence of the theory's penetration—but then *Harry Potter* sold equal gazillions, and no one thinks belief in witchcraft is rampant.

Still, in fundamentalist circles, the Rapture is a given, and comforting it must be, because this is the time in which each Christian will receive his or her resurrected body and "shall be caught up together . . . in the clouds, to meet the Lord in the air." For everyone else, unbelievers, infidels, the unshriven, and the unsaved, tough luck. For them, Christ will return on a horse leading an army, who will exterminate one-third of the Earth's population in a massive act of genocide.

There's one nice new technological wrinkle to all this. In October 2004 the U.S. Food and Drug Administration approved for human use a small rice-grain–sized device called an RFID (radio frequency identification device) produced by a firm named VeriChip, allowing it to be used for human implantation (it was already commonly used as an anti-theft device, in bookstores, for example). VeriChips were intended for use in a number of health-care applications, allowing medical personnel to monitor patients at risk at a distance, through telemetry. Each VeriChip contains a unique ID number and can be tracked through the proprietary Global VeriChip Subscriber Registry. The technology also has obvious security implications. And what has all this to do with the Rapture? One of the precursor events is Satan's marking of every human with the Mark of the Beast or, as Revelation 13:16–18 puts it,

> And [the Antichrist] causeth all, both small and great, rich and poor, free and bond, to receive a mark in their right hand, or in their foreheads: And that no man might buy or sell, save [except] he that had the mark, or the name of the beast, or the number of his name; Here is wisdom. Let him that hath understanding count the number of the beast: for it is the number of a man; and his number is six hundred threescore and six.

The modern Catholic church has interpreted this mysterious number as a coded reference to the Emperor Nero, through some abstruse reasoning that need not detain us here. For the Rapture folk, it is not really much of a stretch to conflate the number 666, Satan's number, with VeriChip, although it represents something of a public relations blow to corporate marketing strategies.[4]

Once you start looking, you can see apocalyptic prophecies everywhere, resurgent now in these days of imminent ecological Armageddon, and as science is uncovering stranger and stranger under-levels to what was once thought to be reality.

Even the newborn atheist Christopher Hitchens, in his polemic against religion and especially against the notion of intelligent design of the universe, expresses his own gloomy view of probable cataclysms:

> [Our] vanity allows us to overlook the implacable fact that, of the other bodies in our own solar system alone, the rest are either far too cold to support anything recognizable as life, or far too hot. The same as it happens is true of our own blue and rounded planetary home, where heat contends with cold to make large tracts of it into useless

wasteland, and where we have come to learn that we live, and have always lived, on a climatic knife edge. Meanwhile the sun is getting ready to explode and devour its dependent planets like some jealous chief or tribal deity.[5]

The end of the world is always nigh. Disappointed seers sometimes fade into oblivion, but most of them regroup. A nice example is the way modern fundamentalist Christian commentators have come to think that Saint Paul was not after all expecting Judgment Day in his own time—he can't really have been wrong, can he? No, he was a saint, one of the biggies, surely infallible. He must really have been talking to the Christians of the twenty-first century, when the End is undoubtedly now scheduled.

Scientists, ancient and modern, are far from immune. Isaac Newton, whose laws of motion are among the most famous in science and who is widely considered one of the great, if peculiar, scientific thinkers of all time, was also an alchemist and a mediocre theologian. In manuscripts recently discovered in the Jewish National and University Library in Jerusalem, the sage discussed his attempts to decode the Bible, which he believed contained God's secret laws for the universe. He also predicted that the second coming of Christ would follow plagues and war and would result in a one-thousand-year reign by saints on Earth—of whom he would of course be one. The most definitive date for doomsday, which he scribbled on a piece of paper, was 2060.[6]

Nor is modernity an antidote to this tendency to see looming disaster. For example, Sir Martin Rees, a former British Astronomer Royal at Greenwich and now a professor at Cambridge, recently published a book with the Revelation-style title of *Our Final Hour: A Scientist's Warning: How Terror, Error, and Environmental Disaster Threaten Humankind's Future in This Century—on Earth and Beyond.* Whew! In it, he predicted the odds for an apocalypse in our time had shortened to 50–50 and were getting worse. How the cataclysm would happen he didn't guess, though he had a few plausible candidates: a genetically engineered pathogen run amok, debris from an erupting super-volcano blocking the sun, scientists accidentally triggering a new Big

Bang as a byproduct of some careless experiment—or maybe nuclear terrorism, deadly viruses, rogue machines, and genetic engineering that could alter human character, "all of [which] could result from innocent error or the action of a single malevolent individual." Among Rees's more modish hazards were nanotechnology ("If the field advances far enough, rogue self-replicating nanotechnology machines, feeding on organic material and spreading like pollen, could devastate a continent within a few days") and particle accelerators ("Perhaps a black hole could form, and then suck in everything around it)."[7]

Of course, science—or rather popular science—has never been shy about predicting disaster. The "lost summer" that followed the eruption of the Tambora volcano in the early nineteenth century was briefly blamed on Benjamin Franklin, who had been experimenting with electrical conductance by lightning rod—the lightning rods must surely have created a conduit for cold air from space.[8] In 1908, there was a brief public panic (well, mostly in America, it is true) that the Earth was about to be swept by a deadly cloud of cyanide gas. It turned out that scientists had found cyanogens, a corrosive poison, in a fly-by comet called Morehouse. Put together with the fact that Halley's comet was due to return to Earth's neighborhood two years later—and that its tail was thirty million miles long and that it would pass by within fourteen million miles of Earth—and there you are: everyone would die.

As recently as 1958 there were media speculations, *pace* the blaming of Ben Franklin and his lightning rods, that space probes, which blasted holes in the atmosphere, were letting in cold air from space.

Somehow astronomers and cosmologists, whose life's work is to peer into the deep heavens and its mysteries and to probe the Beginning, can be forgiven for their doomsday speculations. But these days chemists, biologists, Earth scientists, and even engineers get into the act. In the 1960s, in the still-paranoid days of the Cold War, a Soviet chemist named Nikolai Fedyakin found that water he had encased in thin glass tubes was mysteriously changing, transforming itself, metastasizing into an entirely new viscous form. His results were "confirmed" by an eminent member of the Academy of Sciences of the U.S.S.R., Boris Derjaguin. The Americans, not to be outdone, were soon also finding anomalous water, and so did scientists across Europe. It was called polywater, the presumed mechanism being that ordinary water molecules were somehow becoming chained together as a polymer. It didn't take long before speculations seeped into technical journals as well as the popular press that this polywater might spread, converting all Earth's free water and ending life as we know it. I was in the Soviet Union soon afterwards and remember published pieces there suggesting it had all been a spoof to fool the Americans, but this was just an after-the-fact rationalization, for the early reports were deadly serious. I was reminded of polywater by Malcolm Brown of *The New York Times*, who pointed out that the scare was remarkably similar to a novel by Kurt Vonnegut, whose "Ice9" destroyed the world by the elimination of all liquid water.[9] In the end, the end wasn't so nigh after all, and polywater turned out to be just not-very-clean ordinary water.

The inventions of nuclear physics, and the unraveling of the intricate structures of the universe first by relativity theorists, then by particle physicists and now by string theorists, have also liberated new, stranger, even more terrifying, and horribly plausible speculations about doom. Even before the first nuclear explosion was set off in New Mexico in 1945, there was speculation that the temperatures produced would be so high they would ignite the atmosphere, destroying all life. Not just lay speculation, either—some of it came from Edward Teller, the "father of the H-bomb," who rather stunned his colleagues by speculating that very thing, forcing project director Robert Oppenheimer to commission an investigation. Enrico Fermi, among others, was diverted from his war work to produce reassuring calculations that this surely wouldn't happen. Even so, there was some relief after the first hydrogen bomb went off without cosmic destruction.

Particle accelerators, constructed to boost elementary particles to speeds approaching the velocity of light itself, have become another source of worry. The energies released by colliding protons and antiprotons, for example, approach two trillion volts, with associated temperatures exceeding those inside the sun. Might these titanic energies not tear aside the fabric of space-time itself? If our universe was conceived as a "singularity," and if singularities did exist as theorized inside black holes, and if these singularities were somehow gateways to other universes, as string theory suggested, might not the energy from our universe leak into some other space, perhaps one where there was an absolute absence of energy, an action that would at once annihilate all matter in our space?[10]

Such speculations are still with us, getting ever more arcane as our knowledge of the strange world of quantum physics deepens. In *Scientific American* late in 2005, a physicist named Walter Wagner could be found speculating that the new particle accelerator at the Brookhaven National Laboratory on Long Island (the Relativistic Heavy Ion Collider, or RHIC) would create an entirely new kind of quark-gluon plasma. RHIC is an underground raceway 1.6 miles in circumference, and the object was to accelerate particles to 99.9 percent of the speed of light, at which state collisions would produce temperatures, for a trillionth of a second only, approaching 18 trillion degrees Fahrenheit and would register some forty trillion electron volts. The idea was to recreate for an instant the conditions that would have obtained just after the original Big Bang. Nothing had ever been tried at this scale and at these speeds before—the particles that collided wouldn't even "realize" they had been destroyed, because time would move more slowly than they did. The scientists conducting the experiment didn't really know what to expect, except that this new plasma would contain quarks and gluons, and perhaps other unknown particles, none of which they had been able to study directly before. They particularly wanted to find something called the Higgs boson, without which the Standard Model of the cosmos apparently doesn't work.

Wagner, for his part, suggested that the collisions might just create a new form of matter called strangelets, which would devour ordinary matter. RHIC, he said, may contain within it a strong possibility of causing (a) a transition to a lower vacuum state that propagates outwards from its source at the speed of light, thus consuming the Earth in a millisecond, or (b) a black hole or "gravitational singularity," which would consume the Earth and the solar system in more or less that same millisecond, or (c) create "strangelets," which would—big surprise here—essentially do the same thing. Wagner followed up his article with a lawsuit attempting to stop the experiments before they started, but did not prevail. Then Britain's *Sunday Times* newspaper briefly inflamed things even more by running a banner headline: "Big Bang Machine Could Destroy Earth."

Unnervingly, the laboratory responded that Wagner's calculations had been flawed, but conceded that the possibilities were "not zero." "If strangelets exist (which is conceivable), and if they form reasonably stable lumps (which is unlikely), and if they are negatively charged (though the theory strongly favours positive charges); and if tiny strangelets can be created at RHIC (which is exceedingly unlikely), then there just might be a problem."

Well, it was the science fiction writer Arthur C. Clarke who once suggested that the novas and supernovas we see in the sky might not be ordinarily exploding stars after all. "They might be industrial accidents," he said, some alien RHIC-equivalent consuming their world.

It's easy to scoff. But there have been real catastrophes in human history, and some of them have been encoded in myth and legend. Kevin Krajick, in the journal *Science*, reported on the budding discipline of geomythology and has tracked several instances where real earthquakes and similar phenomena were encoded in folklore long before scientists discovered their on-the-ground traces. One of the examples he found was in Seattle, where in 1990 scientists discovered under the city a previously unknown fault that had ruptured about 1,100 years ago, producing a substantial earthquake that would have caused massive destruction in the modern city; the city's infrastructure was reinforced as a consequence. This quake had been recorded in Native legends.

Pre-Columbian peoples along the Pacific coast of South America once wore amulets, made from shells of the red spiny oyster, that they believed would protect them from adverse weather. Again, there was a link to real events: the species dies off as water temperatures rise, and so spiny oysters washing up on the beaches meant that heavy El Niño rainfall was imminent.

Krajick quotes volcanologist Floyd McCoy, of the University of Hawaii, Manoa, as saying that "[until recently] discussing myth was a good way to sink your credibility," somewhere up there with confessing yourself an Atlantologist. "But I'd be a fool to write it all off. There is a new realization that some myths have something to say."

"Myths can enrich the record," Krajick writes. "Paleo-seismologist Brian Atwater, of the U.S. Geological Survey in Seattle, who has done many studies of seismic events in the northwest, says that legends can reinforce the fact that people do lie in harm's way. The trick is teasing out which myths carry kernels of truth that can be connected to data."[11]

And too much scoffing can obscure the stark fact that real calamities are possible. It wasn't an invention that the asteroid Apophis narrowly missed colliding with Earth in 2004. It isn't a fabrication that massive volcanic eruptions can change the planetary climate in an eyeblink of geological time (and that the lovely mountains of Yellowstone National Park may be the biggest ticking time bomb of them all); it isn't just imagination to say that a million-casualty earthquake is now quite probable (and that even cities like Chicago, far from fault lines, are vulnerable); it isn't a fantasy that a methane eruption could bring much more calamitous global warming and precipitate a sea-level rise of nearly thirty meters. Nor, alas, is it a fiction to assert that we may very well be bringing calamity on ourselves through reckless profligacy and obstinate denial of the facts of global warming. All these, and more, are entirely possible, as we shall see.

So where does that leave us? Before going on to look at what is *out there,* and *in here,* and *underneath the surface*—that is, at our hazardous cosmic neighborhood and our equally hazardous planet—I want to leave you with a little philosophical mummery called the Doomsday Argument. By logical trickery, you can "prove" that the human race is going to die out, soon.

The Doomsday Argument has been the subject of heavyweight philosophical discussion,[12] but I prefer the version by Jim Holt, who describes himself as a "low-voltage journalist who splits his time between New York and Paris [and writes] mainly about philosophical and scientific subjects, occasionally also producing what could charitably be described as humor." In a piece for *Lingua Franca* magazine, Holt (from whom the actual argument set out below is drawn) suggested that "even as transcendental *a priori* arguments go, this one [the Doomsday Argument] is pretty breathtaking. For economy of premise and extravagance of conclusion, it rivals Saint Anselm's derivation of God's existence from the idea of perfection, and Donald Davidson's proof that most of what we believe must be true or else our words would not refer to the right things."[13]

Holt led me to Brandon Carter, the popularizer of the Anthropic Principle, which says (if I understand it correctly) that the laws of physics are what they are because we couldn't have evolved to study them otherwise, and that elsewhere—in other universes, or other pockets of this one—they may be quite different. Carter, in turn, led me to John Leslie, a fellow of the Royal Society of Canada and professor emeritus at the University of Guelph, in Ontario. He's a metaphysician who has been wrestling with, among other things, the neo-platonist idea that God, instead of being an omnipotent creator, is a need for the world to exist, a need that is itself creatively realized, or, alternatively, the notion that the world itself is just the manifestation of the thoughts of a divine mind.[14] Leslie, as you might guess, is a strong defender of the Doomsday Argument.

In any case, the Argument goes like this: if you assume that the human race will survive millions more years, perhaps for the remaining lifetime of our sun, say 5 billion years or so, and that the population of the Earth stabilizes at around 15 billion at any one time, then there would have been at the end of all that about 500 quadrillion humans. Since, at the most, 40 billion or so people have lived on Planet Earth to now, that means that we, you and I, would be among the first 0.00001 percent of all humans. In probability theory (using Bayes's theorem, which essentially says that a hypothesis is confirmed by any body of data that its truth renders probable[15]), the chances of so unlikely an outcome are vanishingly small—ask any gambler. What makes us so lucky, or so special? On the other hand, suppose that humans are wiped out by some catastrophe in the next decade or so. That would make us 40 billionth out of a total human population of maybe 50 billion, much better odds, and therefore much more probable. Conclusion: scenario two is more likely to be true. Therefore: doom sooner rather than later.[16]

If you think this is the sheerest piffle, that it is a position, as the psychologist Paul Bloom said in another context, "that is so intuitively outlandish that nobody but a philosopher could take it seriously,"[17] consider the same line of thinking from another angle. The argument would tell you that the mere fact that the Earth has survived for nearly five billion years does not at all mean that planet-sterilizing disasters are unlikely. Why not? Because observers are, by definition, in places that have avoided destruction. The assumption that catastrophic events are rare fails to take this observation-selection effect into account. We are precluded from observing anything other than that we have survived to the point where we are doing the observing. "If it takes 4.6 gigayears for intelligent observers to arise, then the mere observation that Earth has survived for this duration cannot even give us grounds for rejecting with confidence the hypothesis that the average cosmic neighborhood is typically sterilized, say, every 1,000 years. The observation-selection effect guarantees that we find ourselves in a lucky situation, no matter how frequent such events."[18]

We're lucky, then. Cling to that. That's the good news.

TWO

Catastrophe in Human Life: The Probability Theorem

A chance event is uninfluenced by the events which have gone before.
—Gamblers Anonymous axiom

B ut . . . how real is all this? How probable is disaster? I don't mean the supernatural aspects, or the invocation of gods, or the Doomsday Argument itself, but simply the reality of calamity in the natural world. Is our planet so unstable that it inevitably inculcated a hard-wired fearfulness in our species?

There is, in fact, considerable evidence from many scientific disciplines that the world is not nearly as placid a place as geological Uniformitarianism, the orthodoxy until fairly recently, would have us believe. It is comforting to think that all the changes in the planet, the rising of mountains and the openings of the seas, happened over endless geologic time one tiny increment at a time (mountains worn down bit by bit by erosion, canyons carved out by the endless dripping of water), but it ain't necessarily so, as this book will amply demonstrate.

There have been (relatively) placid times in planetary history—indeed, we have been passing through one for the last ten thousand years or so, though that may be ending—but also convulsive ones.

Even so, natural calamities don't happen according to a timetable, and they happen infrequently. No need to call your insurance broker just yet.

Our vulnerability to calamity is affected by the different time scales on which we and the planet operate. Humankind's entire existence has been a mere nothing in geological time—all the biographies of all *Homo sapiens* are crammed into only a few milliseconds as the cosmos reckons these things. If you look at it positively, the very existence and survival of our species is actually due to that great disparity, the gap between major disasters and the minuscule human lifespan. So when calamity comes, it has tended to come as a terrible surprise. So very long has it been since the last such calamity that the memories have faded, perhaps altogether, perhaps transmuted into legend.

Evolutionists used to wonder why there seemed to be epochs in which nothing very much changed, punctuated by explosive bursts of very rapid evolution. They came to call it the punctuated evolution theory—periods of slow development interrupted by wholesale extinctions and recoveries. Now we know that catastrophism, recurring natural calamities, is almost certainly the reason; volcanoes, earthquakes, and impacting asteroids or comets almost certainly the mechanisms.

This works on a shorter time scale too. One of the great puzzles of recorded history—the era of the written word—is why great civilizations seem so often to have died out abruptly. At least five times during the past six thousand years or so major environmental calamities undermined and destroyed urban cultures around the world.

Some of these, very likely, were due as much to human folly as to natural calamity. The wiping out of the Easter Island civilization has been adduced as evidence in chief; Jared Diamond's view, in his book *Collapse*, is that the island's ecological degradation was caused by overpopulation and the careless stripping of finite natural resources. But there are other examples for which the evidence is more ambiguous. The demise of the much earlier Clovis culture in North America (circa fifteen thousand years ago), and its successor, the Anasazi, was probably due to climate change and consequent desertification, as well as to human misuse of the environment. Other collapses seem to have been caused by climate change too, not necessarily catastrophic in its time span, but deadly to unprepared civilizations. The Akkad culture, whose members built the world's first great cities on the banks of the Euphrates forty-five or so centuries ago, abandoned their dwellings and disappeared, for reasons now thought to have been connected with a severe and prolonged drought; archeological evidence suggests that even earthworms disappeared—a kind of canary-in-the-mine of desertification. The same incident brought about the final collapse of Egypt's Old Kingdom, the preeminent builders of antiquity. The disappearance of the mid-Niger River civilizations around the start of the Christian era coincided with a deeply arid phase in the Sahara. There, too, paleogeologists have been able to trace the growing aridity by following the traces of retreating earthworms, slowly dying out as the desert dried.

Even earlier, around 6200 BC, a severe global cold period devastated farmers in Mesopotamia and the fertile Euphrates delta. There were other such episodes in 3200 BC, 1628 BC, and 1159 BC. These, too, were widely thought to be caused by rapid climate change, the cause *du jour* of our times. But what caused the climate to change that quickly?

At least some of those episodes, it now appears, were caused by natural calamities. Sudden ocean cooling almost certainly precipitated the desertification that ended the reign of the Clovis hunters (and their main prey, the mammoths); the ocean's cooling, in turn, was precipitated by infusions of global meltwater, but it was plausibly also caused by the impact of cometary debris onto the ice pack around where the Great Lakes are now—there will be more on this later. The sudden cold period in 6200 BC was also plausibly caused by the sudden collapse of an ice dam, called Lake Agassiz, where the Dakotas and Manitoba are now, filled originally by glacial meltwater, into the North Atlantic, abruptly changing the ocean circulation and drastically cooling the mid-latitudes in the path of the prevailing westerly winds. The abandonment of the Akkad capital, Tell Leilan in modern Syria, was most probably caused by global cooling that followed a major volcanic eruption, that cooling in turn caused by clouds of debris in the high atmosphere and stratosphere, screening out warming sunlight for years, and possibly decades.

We have better evidence for more recent events. Sometime in the years 535 to 536, well within the reach of the written historical record, something dramatic, and globally catastrophic, happened. Not just the Plague of Justinian, though that was bad enough, as we shall see, and was probably itself facilitated in its spread by the other event.

Whatever it was, a two-year winter followed on its heels. Trees from western America to Europe to Siberia to northern China stopped growing. Crops failed and, inevitably, famine spread. In the subsequent few decades, city-states on every continent collapsed. The event marked the end of Justinian's brave attempt to re-establish Rome's preeminence in what he called Constantinople. The Persian empire collapsed and was later overrun by the Muslims, and the Baghdad Caliphate was born. The American metropolis of Teotihuacán disappeared, the Mayan city of Tikal was abandoned, the Nazcas of South America vanished. As David Keys put it in his book *Catastrophe: The Day the Sun Went Out* (later more grandly subtitled *An Investigation into the Origins of the Modern World*), "It was also the [period] that witnessed the birth, or in some cases the conception of Islam, France, Spain, England, Ireland, Japan, Korea, Indonesia, Cambodia, and the power of the Turks. It also produced a united China."[1]

The Syrian bishop John of Ephesus and the Roman historian Procopius, both of whom left detailed accounts of the ravages of the bubonic plague that horrify even today, described the natural catastrophe too. Wrote the bishop: "There was a sign from the sun, the like of which had never been seen and reported before. The sun became dark, and its darkness lasted for eighteen months. Each day it shone for about four hours only, and still this light was only a feeble shadow. Everyone declared that the sun would never recover its full light again." Procopius compared the sun to the moon, giving forth its light without brightness. Another Roman, Flavius Cassiodorus, marveled "to see no shadow of our bodies at noon. We have summer without heat." And a Chinese chronicler said yellow dust rained like snow.[2]

So what happened? The options include an impact from an asteroid or comet—a notion that until sometime between 1960 and 1980 was thought to be a subject fit only for writers of speculative fiction—or a massive volcanic eruption. But where did this eruption take place, and why don't we know more about it? It can't have been in the Mediterranean basin, or we would have had detailed descriptions. There is no evidence of such an eruption occurring in the Americas of the period, nor in the high Himalayas, the other global hotspots. Which leaves the volcanic "Ring of Fire" around the Pacific. Keys points out that ice cores from both Greenland and Antarctica show elevated acid concentrations somewhere around 527 to 530, a little early but, with the given uncertainty range, plausibly related. Acid rain followed in both hemispheres, which implies that the eruption, if that's what it was, occurred on or near the equator, where the debris field and dust clouds could have caught both prevailing wind systems. His candidate: in or around present-day Sumatra, close to the notorious volcano we call Krakatoa. Perhaps, he suggested, Java and Sumatra were once one island, sundered by the cataclysm.[3]

It is possible, even likely, that the human species itself came close to extinction through just such a natural catastrophe.

The catastrophe itself has left enough traces that volcanologists are confident both of its date and its effects: when Toba in Sumatra blew up 75,000 years ago, it was the biggest mega-volcano eruption in the last 2 million years, hurling more than 120 cubic miles of debris high into the atmosphere, perhaps as high as 19 miles, leaving its traces over 1.5 million square miles, about half the size of the continental United States. Ash from the eruption has been recovered from deep-sea cores taken in the Bay of Bengal and even deep into India, more than 1,900 miles away. On Samosir Island, close to the eruptive source, the volcanic tuff is more than 2,000 feet thick. The caldera (the cavity left after the eruption) is 19 miles long by 62 wide. By way of comparison, the Mount St. Helens eruption of 1980 ejected only 0.6 mile of rock and ash; the cataclysm 600,000 years ago at what is now Yellowstone National Park in the United States—a cataclysm that stripped half the continent of all life—was perhaps a little smaller than Toba.

The effects on living creatures of such an eruption would have been catastrophic. No plants, animals, or humans in the region would have survived in the immediate aftermath. But the effects were more widespread than that, and much longer lived. These mega-eruptions have many effects not just on the biosphere but on climate itself. The sulfur dioxide in the volcanic cloud combines with water vapor to form sulfuric acid, which not only causes ruinous acid rain but scatters, reflects, and absorbs sunlight, drastically cooling the surface and reducing photosynthesis. Global temperatures would have plunged somewhere between 41 and 54 degrees, causing a six-year-long volcanic winter and causing mass die-offs everywhere. It wasn't a short-lived cataclysm: we know that the last glacial period was preceded by almost a thousand years of the coldest temperatures of the Late Pleistocene era.

All of which may explain something that has puzzled evolutionary theorists for some time—somewhere around seventy-five thousand years ago, just at the time of Toba's eruption, there seems to have been a bottleneck in human genetic diversity, based on studies of mitochondrial DNA distribution patterns. Before the Toba episode, humans were diverging rapidly ("modern" humans reaching as far as India), during it they were squeezed to a few DNA lines, and afterwards they diverged again. Beguiling (if contested) genetic evidence suggests that all humans alive today, despite our apparent variety, are descended from a very small population, perhaps between a thousand and ten thousand individuals—a single concert hall can hold more people than that. It is clearly not possible to prove cause and effect, but the coincidence suggests something dramatic happened.

Stanley Ambrose, a paleoecologist at the University of Illinois (he is director of the Environmental Isotope Paleobiogeochemistry Laboratory) and a specialist in African pre-history, thinks he knows what it is: "Genetic evidence suggests that human population size fell to about 10,000 adults between 50 and 100 thousand years ago," he wrote in the *Journal of Human Evolution*.

> The survivors from this global catastrophe would have found refuge in isolated tropical pockets, mainly in equatorial Africa. . . . Populations living in Europe and northern China would have been completely eliminated by the reduction of the summer temperatures by as much as 12 degrees [Celsius]. Volcanic winter and instant ice age

may help resolve the central but unstated paradox of the recent African origin of humankind: if we are all so recently "out of Africa," why do we not all look more African? Because the volcanic winter and instant ice age would have reduced population levels low enough for founder effects, genetic drift and local adaptations to produce rapid changes in the surviving populations, causing the peoples of the world to look so different today.

Toba, he suggested in a more speculative vein, may have promoted the evolution of social cooperation that facilitated dispersal out of Africa, causing modern differentiation abruptly rather than gradually over a million years.

It's fair to say that Ambrose's theory is not without its critics—and supporters. Among the supporters is Michael Rampino, an earth scientist at New York University, who agrees that Toba would have set off a major cooling, with at least two years of some equivalent to a "nuclear winter." The critics include F.J. Gathorne-Hardy, who toils for the Natural History Museum in London, and W.E.H. Harcourt, a paleontologist at the American Museum of Natural History. In a trenchant paper published in the *Journal of Human Evolution*, they disputed the bottleneck theory, suggesting that the shrinkage in population—they don't deny a shrinkage took place—happened over a much longer period. They also point out that there is scant evidence that any other species went extinct at the time, though almost all animal species are far less adaptable than humans, which in their minds casts doubt on the notion that human populations were affected worldwide.

That a major calamity happened, though, is not in dispute. Nor is the fact that our species was grievously affected.[4]

We're the very product, then, of calamity. It made us, shaped us, nearly killed us, and then sent us packing, one more time, out of Africa, to where we are today.

Now our attention is being torn from one disaster to another. Over the past decade, there has been an average of one natural disaster per day—348 recorded events each year. This number comes from the Centre for Research on the Epidemiology of Disasters, which defines a disaster as a natural hazard that must kill at least ten people, affect one hundred or more, and necessitate a declaration of emergency or a call for international assistance. Disasters included are drought, earthquake, extreme temperature, flood, slides, volcanoes, waves/surges, wildfires, and windstorms. A calamity, in the hierarchy of things, is greater than a disaster. A calamity will kill or affect at least a village, and probably a region, with no upper limit.

Can such calamities be predicted and therefore at least to some extent planned for? What are the probabilities that a hurricane or a tsunami will hit your community, or that a killer volcano will erupt, an earthquake shake the country, an asteroid hit the planet, a pandemic wipe out millions of people? Probability theory can't tell you when such an event will happen—but can it tell you the *likelihood* of such an event?

The answer is either, sort of, or not really.

It has to do both with the large number of variables at play and the difficulty of making credible assumptions.

The law of large numbers in probability theory goes something like this: "If the probability of a given outcome is P, and the event is repeated N times, then the larger N becomes the likelihood increases that, in proportion, the outcome will be N*P." So if the probability of throwing a double six with two dice is 1 in 36, then the greater the number of times you throw, the closer the outcome will be to 1/36. The crucial point is that the number of successful throws goes up—but the number of unsuccessful throws goes up faster, in the proportion 36 to 1. Ordinary people call this the law of averages. The "gambler's fallacy" is a misreading of the law of averages. Gamblers like to believe that, in the long run, things will even out—forgetting that crucial phrase "in proportion." If a coin is spun a hundred times and has landed heads sixty times, gamblers will often assume that tails are now due for a run to get even. And they are right—but it may take millions of throws before the proportions do even out, long after even the richest gamblers would have been bankrupted by the house. The reality is this: as the number of tosses gets bigger, the probability of the percentage of heads to tails gets nearer to 50 percent—but the difference between the actual number of heads or tails thrown and the number representing 50 percent also gets larger.

An egregious example of the gambler's fallacy at work is from the Italian lottery. In 2003, someone noticed that the number 53 had not occurred as a winning number for more than two years (the lottery invites punters to bet any amount on any number between 1 and 90 every two weeks). During the following two years, betting on 53 became a national obsession, as punters were convinced its "turn" must come soon. Several billion euros later, its failure to appear was blamed for several deaths and bankruptcies—early in 2005, a woman drowned herself in the sea off Tuscany after she bet her family's life savings on number 53; a man living near Florence shot his wife and son and then himself because of his number 53 debts. When the number finally did appear late in 2005, a consumer group that had called for its banning "to halt the country's psychosis" said it was delighted that the national bewitchment had been broken.[5]

Obviously enough, it would be much easier to predict the throw of dice, or the turning up of a specific number in a lottery, or the arrival of a rogue asteroid, if we had more information than merely a theory of probabilities. Loaded dice make an easy example—loading the dice changes the probabilities in a knowable way. Insider trading on stocks is another example, which is one reason the practice is generally banned. Others are more complex: the probability that a suspicious spot will be a tumor is greatly affected by the patient's family history, age, previous medical history, and so on. Natural calamities are like this. To some degree they are random, or stochastic, events—we can sometimes see why they happen, and often can see where, but seeing the precise "when" is very difficult. We still suffer from information scarcity. We can "load the dice," at least a little, but the number of variables is so large, and many of them are so unquantifiable by current techniques, that real predictions remain impossible.[6]

In just this way, probability theory can tell you the chances of an asteroid hitting Earth—but it can't make any sensible predictions about when, or how often, it will happen without knowing a good deal more about (a) the actual number of asteroids in near-Earth orbits, (b) their composition, (c) their velocity and trajectories,

(d) their size, and (e) whether their orbits fit collectively into some kind of organizational pattern. Even then, the amount of residual uncertainty would be very great. If the asteroid orbits themselves are entirely random, then the theory can tell you, essentially, nothing. Finding them, and watching them, is the best you can do, and a much better strategy.

Similarly, theory cannot tell you the probability of a particular volcano erupting, without detailed information on a huge array of variables, such as the size and character of the underlying magma chamber, the structure of the rock through which it vents, the mixture of gases and minerals of which it is composed (which may change over time as the underlying mantle convection cells move), and dozens of others. Geologists can tell whether the volcano is active or inactive, whether it has erupted before, and with what periodicity, but even then, without the other variables being known, no prediction about the next eruption is possible.

The same is true of hurricanes. We know they happen every year, and we know more or less where and how. But no matter how hard we try, we can't know exactly when and why—tiny changes in the environment in which hurricanes are conceived can make the difference between an embryo cyclone expiring and one maturing into a severe hurricane. What these tiny changes are, we don't know. All we know is that they are too small for our current technology to detect.

This is not at all to say we are without any predictive abilities. We are pretty sure which regions will suffer from hurricanes and typhoons, and roughly when and where. We know which regions of the planet are seismically more active than others, enough to suggest that earthquakes are more probable here rather than there. And NASA, as we shall see, has been tasked with keeping track of nearby asteroids so we can zap them before they become a menace. Analyzing effects, understanding systems, and reading history can speak of likelihoods in another way.

In 2005, for example, a team led by Shinji Toda of the Geological Survey of Japan re-evaluated the seismic risks to Tokyo in light of past earthquakes. In a report to the government, they found that a modern replay of the 1855 magnitude 7.3 Tokyo earthquake (an event that killed seven thousand people) could devastate the city, cause trillions of dollars in damage, and send shock waves around the world's financial markets. Tokyo is particularly vulnerable, because it lies at the junction of three separate tectonic plates—which also, the report admitted, means that the uncertainty factor is great. An earthquake will happen. But exactly how severe it will be, and when it will come, can't be known. It could happen tomorrow. Or in a hundred years. There is no way of telling.

Other places on the planet are more worrying even than Tokyo. The most dire possibilities lie in the known earthquake belts across Asia, many of them containing rapidly growing cities filled with unstable buildings, making a million-fatality earthquake probable in the near future. A report in the journal *Nature* quoted an expert as predicting that Istanbul, Tehran, and Sumatra's Padang were likely sites. "Tehran, a city of 12 million, with what one expert calls some of the scariest construction known to man, is a particular worry. . . ."[7] But as with Tokyo—when such an event will occur remains a mystery.

And volcanoes? The twentieth century's largest volcanoes were far smaller than the monsters of the nineteenth, like Tambora (1815) and Krakatoa (1883), both of which affected global climates for years. Even Krakatoa was an order of magnitude smaller than the Santorini eruption of 1400, which wiped out the Minoan civilization. Which means what? Does this diminution make it more or less likely that a truly massive eruption will occur? On this, we have no information at all.

Which leaves us where? Given enough time, and enough people, individually improbable events become increasingly likely to happen. Some of them will be calamitous, but we don't know how calamitous. Some of them might not happen for decades, or centuries, but happen they will. Many people will die, though we don't how exactly how or when. But most people, most of the time, from most of the calamities, will survive. That's the rest of the good news.

PART TWO
Context

THREE

Our Perilous Neighborhood: Understanding Cosmology

Things do go whizzing about.
—Bertie Wooster, in *Leave It to Jeeves*

The feckless Wooster of P.G. Wodehouse was talking about mice in the basement and worrying more about his ankles than the cosmos, but he spoke truer than he knew: whizzing about has been built into the structure of the universe from the very beginning. Restlessness is the governing condition of all matter, and the notion that "space" is empty and that eternal silence is all that is found in the intergalactic interstices is coming more and more to be seen as quaintly old-fashioned.

Violence is endemic to the universe, its only real constant. Our world was begat in violence, and it can be plausibly argued that the period of calm in which the human species was formed is only a brief drawing of the breath before the cosmic assault begins once more.

The theory of the origins of the universe goes something like this: at one instant there was nothing—no matter, no time, no space,

31

no galaxies, no dark matter, no Earth, no video games or politicians or spin-meisters or errant teenagers—nothing. Then suddenly, 13.7 billion years ago, there was something—an infant universe of expanding gases. This was the Big Bang: in a billionth of a trillionth of a trillionth of a second this new . . . *thing* . . . doubled and redoubled a hundred times, until each atom was the size of a galaxy. This is called, rather anticlimactically, inflation.

This theory pays no attention to the "why" of this titanic explosion, only to the "when," the "what," and the "how." The why, which is to say "why is there something rather than nothing," is regarded as an essentially theological question, that place where science and the supernatural come together in mutual puzzlement, and therefore not worthy of further investigation.

The other great unanswered question (which also trespasses into theology) is this: what came before the Big Bang "singularity," as scientists call it?

Most conventional post-Einsteinian science asserted that the question was meaningless. Since time itself began at the singularity, there could not have been a "before" to be, well, before. . . . There wasn't even nothing—there was just not anything. The Copenhagen Interpretation, on which this classical view is based, merely said that the world is decided when the many possibilities of the quantum world "collapse" to become the certainty of the classical or physical one. Which, if you can get around the peculiar scientific usage of the word *collapse*, seems to imply that the world is made up as we go along. More recently, theorists have suggested that our universe is simply a quantum fluctuation in some pre-existing region of space-time, and that this can work both ways—ours came from some earlier universe, and ours leads in turn to further universes, in a process that really has no beginning and no end. This may be equally baffling, but it does get around that counterintuitive notion that something started up without there being anything to actually start it. One version of this, the baby universe theory, suggests that the spawning process takes place in black holes, themselves "singularities," and that a collapsing black hole emerges in some other space-time entirely. In other words, the Big Bang was a transition, rather than a beginning.

Another notion, the many worlds hypothesis, was first floated by a young physicist from Princeton, Hugh Everett, in 1957. Everett had been puzzled—and irritated—by one of the conundrums of quantum physics, which is the knowledge that quantum particles evolve into a weird state called "superposition" in which, among other characteristics, they can be in two places at once. Physicists have always been somewhat unsettled by this idea, because it cannot be observed—if particles can achieve superposition, why can't, say, a dining room table? Or we ourselves? Everett's solution was that such objects *are* superposed, only you can't see them because you yourself are part of them: "When we encounter an object of superposition of, say, here and there, that superposition draws us in too; splitting us into one being who sees the object here, and another who sees it there." The implication was that the universe perpetually splits into many worlds, all of them coexisting side by side, an infinity of parallel universes.[1] Everett's notion is belatedly gaining substantial support among cosmologists.

No one knows when the universe will end, either, or even if it will end. Until recently, cosmological theory suggested that our universe, at least, was only a small way through its probable existence, from the Big Bang to the Big Squeeze, but the theory also suggested that our universe is not as stable as we would like to think, but only "metastable," that is, that it could slip into a different energy state at any time. If this were to happen, it would cause a little bubble of zero energy and true vacuum to appear, which would then annihilate everything—us, our planet, our star, our galaxy, all distant galaxies, and possibly the universe itself. Too bad for us, but the universe of universes would hardly notice—there may be an infinity of universes, after all. Now, the theory that the universe would collapse into some big squeeze has been discarded. To everyone's astonishment, measurements have shown that the universe's expansion is not slowing down at all, but actually speeding up.

In any case, at the moment of the particular singularity that created our particular universe, all matter and energy were concentrated into a single point of infinite density, somewhere around 14 or so billion years ago. The Bang itself exploded outward from a point smaller even than a proton, into the observable universe we see today, filling this nascent universe with cosmic radiation (called CBR, or Cosmic Background Radiation, which, curiously, can still be heard—a tiny percentage of the hissing static you hear when your radio is mistuned is sound that is 13.7 billion years old).

A few nanoseconds after the Big Bang—the theory goes—the distribution of matter and energy in the infant universe was almost perfectly smooth, but small perturbations occurred almost instantly. Experiments have measured tiny irregularities, a tiny fraction of a tiny fraction of 1 percent, in the intensity of the brightness of the background cosmic radiation, the vast sea of primordial stuff that was the universe after a mere few hundred thousands of years. What caused these irregularities is of course unknown, but they do show that entropy—the notion of inbuilt disorder—was operating from the very beginning. In time, and under the influence of gravity, these tiny fluctuations gradually coalesced into a web of hydrogen gas and "dark matter" from which stars and galaxies and everything else would later form.

This question of dark matter, as well as its companion dark energy, remains contentious. We don't know what it is, we can't see it or measure it. It may not be there at all—it is only there in the theories because they just don't work otherwise. The textbooks will tell you that the nature of "dark matter" remains mysterious; the wiser books usually hedge even that bet and admit that the very existence of this dark matter remains in dispute. In July 2007, *Science* reported that dark matter had in fact been formally detected, by measuring the gravitational deflection of light from distant stars. This "map" revealed a ring of dark matter 2.6 million light-years across, surrounding the galaxies under study. Even so, no one could explain what it actually was, except to say that dark matter, if it is dark and if it is matter, provides a sort of super-scaffolding around which all the other structures in the universe, from galaxies to superclusters, have taken shape. A month later the same journal published a speculative piece by Tracy McGaugh of the University of Maryland, suggesting that what was thought to be dark matter might after all just be a slightly modified gravitational force, as predicted by something called Modified Newtonian Dynamics. . . . So it might not exist after all.

Whatever it is or isn't, it does have mass. The universe is now thought to be composed of 4 percent ordinary matter (the stuff we stand on, eat, and are ourselves composed of), 22 percent dark matter, and 74 percent of something called "dark energy," about which no one knows anything at all except that it somehow stretches space; cosmologists tend to resort to non-scientific language like "weird" when describing what it is thought to do. Leon Lederman, co-author of a seminal 1993 book on high-energy physics, puts it this way: "We haven't a clue as to what it is, but we know what it does: it maintains a continuous outwards push on the matter of the universe, sustaining and increasing the expansion rate, and thereby counteracting the gravitational attraction that should be slowing the expansion [but is not]. It might not be dark, it might not be energy. But it accounts for more than 70 per cent of the mass of the universe."[2] In 2007 the U.S. National Research Council suggested to NASA that exploring dark energy should be given top billing in its so-called Beyond Einstein exploration program.

The primeval hydrogen and dark matter web slowly formed into concentrations that are now called halos; eventually these halos closed in, through gravitational forces, and drew in enough of the primordial gas to build first-generation stars and proto-galaxies.

The presumption is that, in the absence of heavy elements, the cooling of the primordial gas would have led to an early generation of short-lived stars, much hotter, much bigger, and much brighter than anything currently in existence. After a mere few hundred million years, these massive stars went supernova and exploded, hurling huge clouds of heavy elements formed in their thermonuclear interiors into the cosmic gas. This—the very first act of polluting the environment, unless you count the Big Bang itself—dramatically changed the criteria for future star formation, by creating the heavy elements and metals that were missing before.

New stars, made up partly of the heavy-metal pollution detritus, probably formed before coalescing into galaxies, which are essentially accretions of second- or third-generation stars. Which is not to say that galaxies were unimportant. By binding the stars and gas together they created cosmic ecosystems, which were a precondition to the assembly of enough heavy elements to make planets, and therefore life, possible.

Every single tiny living cell in the universe is here because these enormous structures called galaxies came into being.

One more thing is needed to understand the notion that . . . things . . . are lurking out there waiting to do us damage, and that is some sense of the scale of the place we live in.

How far away is the rim of, well, everything?

If the universe has shape, we can't detect it. Most cosmologists assert that there is no such thing as a universal center point anyway. However, hints emerged in 2007 that the whole thing might have an axis, after all—all those billions of galaxies might be rotating around this axis, like giant bits of meat on an endless skewer.

And how big is endless? If the universe is 13.7 or so billion years old, and if nothing can travel faster than light, then the maximum size must be, well, a radius of 13.7 billion light-years, or 27.4 billion in diameter. But it is a great deal more slippery than that. The figure of 13.7 billion years refers only to when the first galaxies started emitting light, and the light started traveling through space. But space itself was expanding as the light made its journey, at first very quickly, and then more slowly. Because of this expansion, it took 13.7 billion years for the light to reach us, and we are only 1 or 2 billion light-years away from its source. Put another way, 13.7 billion light-years is only the maximum distance light could have traveled in the time available and bears no relation whatever to where anything actually is and only a tenuous relation to where anything was. The expansion of space probably puts that same galaxy maybe 30 billion light-years away from us; the edge of the observable universe is probably around 46 billion light-years away, and the size of the universe itself would be double that, or 92 billion light-years. But no one really uses numbers like these. In universal terms, they are more than ordinarily useless, because it is far from certain, in the unlikely eventuality you were to show up billions of light-years from the center, if there is one, that you would run up against some kind of cosmic exit sign. The real answer is that if there is nothing beyond the universe, then you can't get there, can you?

In any case, a "light-year" is just a shorthand way of saying appallingly big. In a year, light can travel just shy of 6.2 trillion miles, which is why astronomers have invented an even larger number, the parsec (which stands for *pa*rallax of 1 arc-*sec*ond), which is calculated at about 3.26 light-years. A parsec weighs in at around 19 trillion miles, or 19 followed by 12 zeroes. A parsec is a pretty long way. I mean, it's already a long way from the Earth to the sun, around 93 million miles, but you'd have to make more than 100,000 round trips to the sun before you'd covered 1 parsec.

You can see why light-years and parsecs are useful in discussing even galactic distances. To reach our nearest galactic neighbor, Andromeda, a couple of million light-years away, you'd have to travel some 13 quintillion miles, or 13 followed by 18 zeroes, and after a while you'd fill up whole pages with zeroes, causing your readers to cringe and the paper makers to gloat. Our solar system would have to be more than two thousand times larger than it is to fill a single parsec.

It's hard to see where in the universe we are—partly because of the awkward fact of the place having no fixed boundaries and having, in effect, an uneven shape. The largest things in the universe ("structures," to use the more technical term) are the superclusters that form a web of more-or-less matter throughout the universe. Looked at this way, we're sort of in interstellar Hicksville, located in a small group of galaxies (the Local Group) toward the periphery of a smallish supercluster named, prosaically enough, the Local Supercluster or sometimes, more poetically, the Virgo Supercluster.

This local group, a kind of universal suburban knitting club, has three large and thirty or so smaller paid-up members. If it makes you feel better in all this immenseness, the Milky Way is the biggest fish in its little pond, more massive than the actually rather larger (in size) Andromeda Galaxy, our nearest large neighbor at 2.9 million light-years away. Other group members are much closer than that; the nearest of all, discovered only in 2003, is the Canis Major Dwarf Galaxy, which is a piffling 25,000 light-years away. This disintegrating dwarf, and another rather prosaically called SagDEG (for Sagittarius Dwarf Elliptical Galaxy), or sometimes Sagittarius Dwarf Spheroidal Galaxy, not to be confused with SagDIG, or Sagittarius Dwarf Irregular Galaxy, a different creature altogether (with so many objects in the universe it pays to be precise), are both crossing our orbit and will in due time be either absorbed or annihilated. A little farther away are the Smaller and Greater Magellanic Clouds, at 210,000 and 179,000 light-years.

Our own neighborhood—the spiral-shaped galaxy we call the Milky Way,[3] a pinwheel of stars and gas loosely bound by gravity—is big enough, probably between 90,000 and 160,000 light-years in diameter, and somewhere around 3,000 light-years thick, shaped like a wispy disk with a slight bulge in the middle. The disk has plenty of spiral arms and a dense nuclear core called, for obvious reasons, the Galactic Center. Of course, there is no defined boundary, so it's hard to know when to reel in the tape measure. Most of the mass lies within 50,000 light-years of the center; we're well within that range, at about 24,000 light-years out. Somewhere between 200 billion and 400 billion other stars and their unknown number of planets exist in the galaxy, along with many millions of globular clusters and nebulae, some of them apparently only recently scooped up by us from dwarf galaxies that blundered into our neighborhood. The galactic mass is probably between 750 billion and 1 trillion "solar masses," using our own sun as a baseline. And if we accept the standard theory about dark matter, there must be much more of it than visible matter— perhaps 5,000 billion star-sized dark objects lurking in the shadows of just our one galaxy.

The Galactic Center consists of old stars more highly concentrated than on the periphery and contains what are known as globular star clusters, about two hundred of them. You can "see" this center, or at least infer it, if you squint out in the direction of the constellation Sagittarius close to the borders of neighbor constellations, Scorpius and Ophiuchus. Because we're on an arm, we see the rest of the galaxy as a flattish plane, which shows up as a luminous band called for convenience's sake the Galactic Equator. Our small arm connects with much bigger arms, the Sagittarius–Carina Arm inwards from us, the Perseus Arm outwards.

Also at the center is a massive black hole, and it is lucky for us that we're a fair distance from it. The center would not be a pleasant neighborhood. It's violent enough around here, as we shall see, but in the center periodic flarings and massive explosions take place, hurling great jets of supercharged particles more than two light-years into space. Great lobes of super-hot gas reach even farther, up to ten light-years away, with temperatures reaching 68 million degrees Fahrenheit. Perhaps predictably, there is not much matter left in the immediate vicinity.

Just as in other galaxies, supernovae occur in the Milky Way from time to time. The last seen from Earth was in 1604 and was studied by the great astronomer Johannes Kepler (when he wasn't tinkering with what must surely be the earliest science fiction yarn, titled "Journey to the Moon in the Arms of a Daemon"). Astronomers keep hoping for another so they can watch it through modern telescopes, all the while hoping that it is close, but not too close.

The Milky Way, then, is an ecosystem just like, although larger than, ecosystems on Earth, with its own laws, birth, life, and expiration—like everything else, it has a best-by date and is fated to expire. It consists of stars, radiation, gas, and planets that interact and intersect in complex ways whose pathways is still guesswork. An ecosystem, but not a closed one. Galaxies rarely evolve in isolation and sometimes merge and collide.

One of the more interesting questions facing cosmologists now is this: is our place, our actual location in our galaxy, crucial to explaining why life formed here? So far, ours is the only known (to us) solar system to contain life. Is where we are in the galaxy a coincidence, or is our location more hospitable to life than it would be either farther in or farther out? As the anthropic principle would put it, we evolved here because here is the only place we could have evolved.

Is there, that is, a galactic temperate zone?

We know planets exist elsewhere. By mid-2007, 236 "exo-planets," planets outside the solar system, had been identified, and more were being discovered every week. One star, a giant called Upsilon Andromedae, has at least three planets. Most exo-planets are massive, closer in size to Jupiter than to Earth; tend to be alarmingly close to their parents; have eccentric orbits; and most orbit red dwarf stars, which though long-lived are probably too cold to support life. But some exo-planets are unexpectedly hot: a little thing unromantically called HD149026b, which is thought to be enriched with heavy elements, was found to have an ambient temperature of 1,741 degrees Kelvin, about five times as hot as boiling water, somewhat more than shirtsleeves weather. A detailed search for planets in the very old and quite central globular cluster called 47 Tucanae turned up no planets at all—perhaps, it is thought, because it is so cluttered with matter that collisions are inevitable, destroying embryo planets before they can really get started. But then, in May 2007, a team of Swiss astronomers claimed to have discovered the first Earth-like rocky planet circling the star Gliese 581, itself a red dwarf, about 20 light-years toward the center. The discovery was disputed, however. Six months later, a NASA-funded study by a team of American astronomers described a solar system similar to our own, with a sun, a star called 55 Cancri in the Cancer constellation, only about 40 light-years from us, of the same size and age as ours. Cancri's first planet was seen as early as 1997 by astronomer Geoff Marcy and his collaborators at UC Berkeley, and a second was spotted in 2002. In November 2007, Marcy's team said that they'd cataloged a further three planets, all of them much bigger than Earth.

If you look at the galaxy as a whole, star formation rates and the abundance of heavy elements decrease the farther you get from the center. It is entirely possible that closer in the stars are too hot, too active, and the radiation too intense, to allow the formation of any kind of life; farther out, there are probably not enough heavy elements, particularly radioactive ones, to sustain life were it ever to spring into being.

Our own sun would have been formed in the familiar way. The process starts in the cold, dense heart of the gas clouds called molecular clouds, whose slow accumulation has brought them to the verge of gravitational collapse. Out of this first-stage collapse come proto-stars, baby objects with widespread gas-rich dusty clouds reaching far into space. These gradually coalesce into main-sequence stars, which are still surrounded by massive clouds of ice and dust that remain after the gas disperses. It is this stellar debris that makes planets.

Indeed, this is where we come in, about five or so billion years ago.

This cosmic dust and ice began to coalesce, and within a few eons formed billions of objects probably 6 miles or so in diameter. Each of these had a tiny gravitational field of its own, and the process of accretion through collision began, with galactic space a kind of cosmic snooker table, with six-mile balls careening into each other—but sometimes, if the velocity were small enough and the angle low enough, sticking together. Once the process began they would grow rapidly.

About a million years after it began, these proto-planets would have been somewhere around the size of Mars or our moon. Once they got to be about 19 miles or so in diameter, gravity would have been substantial enough to sink heavier elements into the core, and they began to heat up, partly through radioactive decay of short-lived radionuclides such as iron, and partly through kinetic energy supplied by the continuing collisions.

By the end of this stage, there would have existed many hundreds of small planetesimals in still irregular orbits about the sun, with still millions, and probably billions, of uncollected fragments. The very irregularity of their orbits set up the next, more cataclysmic, stage of further growth through collision and accretion. This would have lasted somewhere around a billion years.

Our own planet, Earth, born somewhere around four and a half to five billion years ago, was not immune to cataclysm. Somewhere in Earth's early history a small body, probably somewhere around the size of Mars, usually now referred to as Theia, smashed into the planet with enough energy (How many billions of atomic bombs? Maybe a trillion or so . . .) to melt it all the way through. Much of the impacting body was also melted and merged with Earth, but a small world's worth was flung back into the sky, where it hangs yet, causing tides, titillating lovers, and making werewolves itchy.

Geological Ages

Precambrian

Hadean *4500 Ma to 3900 Ma*
Archaen *3900 Ma to 2500 Ma*
Proterozoic *2500 Ma to 540 Ma*

Paleozoic *(ancient life) 540 Ma to 245 Ma*

Cambrian
Ordovician
Silurian
Devonian
Carboniferous
Permian

Mesozoic *245 Ma to 65 Ma*

Triassic
Jurassic
Cretaceous

KT Boundary *(dinosaur extinctions)*

Cenozoic *65 Ma to present*

Tertiary

    Paleocene
    Eocene
    Oligocene
    Miocene
    Pliocene

Quaternary

    Pleistocene (1.8 Ma to 10,000 BC)
    Holocene (10,000 BC to present)

(Ma = million years from the present)

Very likely, the outer layers of Earth melted, deepened, and became shallow many times after the impact, remaining unstable for perhaps another seven hundred million years or so. Scientists call this the Hadean epoch, after the underworld of Greek myth.

The cosmic neighborhood has settled down since then. Many of the loose objects have already been destroyed, absorbed into one or other of the planets, or have disappeared from the sun's grasp. But not all of them. A few billion bits were left over. And they are, as Bertie Wooster put it, still whizzing about.

When you get down to our own solar system, the distances become more manageable. We don't use light-years to measure our small place under the sun. Instead, astronomers use something called the Astronomical Unit, or AU, which is the distance from the Earth to the sun, somewhere around 93 million miles. We're one AU from the sun. Mercury is about a third of an AU, the newly demoted Pluto about 40 AU.

Even so, the numbers are mind-bogglingly large.

The diagrams of the solar system in textbooks and museums are generally quite misleading, and so are those wonderful eighteenth-century mechanical models called orreries (after the Earl of Orrery, who had one built for him in 1715). Even the largest such models, such as the one in the Hayden Planetarium in New York, are far too small and out of scale. Even when they altogether omit the three outer planets, they can't realistically show the others to scale. They are too far apart. Even if you drew the Earth as a tiny thing the size of a peppercorn and Mars the size of a pinhead, Pluto would be more than a kilometer away and too small to see without a microscope. Our nearest neighbor by far is Venus, and it is so very distant that it exerts exactly the same gravitational attraction as a single high-rise building nearly three hundred miles away.[4]

Even Pluto isn't the end of the solar system. That would be the Oort cloud of drifting comets, a thousand times farther. The Oort cloud has never been seen, only deduced. The Dutch astronomer Jan Oort, who was the first to determine that the Milky Way actually rotates around its center, inferred the cloud's existence in the 1950s from the long-period comets that occasionally entered the solar system, measuring their orbits and extrapolating backwards into space.

The Oort cloud is a massive, probably spherical, "cloud" that envelops the planetary system, extending maybe three light-years out into space, or some 19 trillion miles, at the very edge of the sun's gravitational influence. The cloud is thought to contain somewhere around six trillion "objects," mostly icy balls or comets. The comets near its outer edges are only weakly bound to the distant sun and can be readily deflected either by nearby stars or by molecular clouds and galactic tides, which may nudge them from place and send them either deeper into the solar system or out into interstellar space. Oort figured that the mass of the eponymous cloud was some forty times that of Earth. But don't think you can ignore the Oort cloud altogether: this is where comets such as Halley and Swift-Tuttle originate. And they affect us directly.

The point of all this discussion about enormous distances, about violent and unpredictable beginnings, about cataclysmic clashes and gigantic collisions, about insane collapses and mighty explosions, is not to beat the reader into an awestruck submission, but just to recognize that cataclysm is not only our universe's normal condition but is an inbuilt part of the process—without this almighty violence, nothing would ever get done. None of these things, whether on the universal scale or the planetary one, are calamitous—except to those conscious beings directly affected. The universe exploded so we could have matter. The early stars exploded so we could have the heavy elements life needs. Supernovae and quasars are a normal part of the cosmic cycle. Planetary collisions were necessary to clean up the neighborhood—a process that still isn't finished, so perhaps we arrived on the scene a little early. And in the case of our own small parcel of the universe, the part we call Earth, the earthquakes and volcanoes are simply a planetary way of maintaining, literally, its cool. It's not the planet's fault we get in the way. But we can learn to live with it.

FOUR

This Plastic Earth: Plate Tectonics and Wandering Continents

Naturally, nothing is safe in this uncertain world.
—Geologist Richard Fortey

Our Land Cruiser left the Mosquito River at dusk and wound steadily up a steep, heavily treed escarpment, skirting huge pot-holes and equally huge piles of elephant dung. By the time we got to the top, it was black night, no moon, thin overcast scudding across the stars. The road followed the crater rim, heading east at first, then curving northward. To the left of the car, I knew, it dropped off sharply to the crater bowl, the ancient volcanic caldera called Ngorongoro, but of that we could see nothing.

By dawn I was sitting on the crater rim with a Maasai named Viola ole Yaile, watching the sun come up, blood-red and massive, over the distant escarpment. In the far distance, out of sight even in daylight, were the Serengeti plains, stretching all the way to Lake Victoria and Rwanda. Beyond the crater, to the east and southeast, were the lush hills of the Great Rift Valley escarpment. In front was the crater itself, left over from some volcanic catastrophe, grazing or hunting grounds for a wild assortment of African fauna, living either in savagery or in some natural and harmonious balance with their environment, depending on your perspective. It was home to some seventy-five thousand animals (including the omnipresent zebra, gazelle, springbok, and wildebeest, and the Big Five—rhinoceros, lion, leopard, elephant, and Cape buffalo).

The other rim was 14 miles distant. Ngorongoro is the largest unbroken and unflooded volcanic caldera on the planet, not as large as, say, Yellowstone in Wyoming but much less fractured. The caldera floor was some 2,000 feet down from where I sat, and very level. Streams ran down into the crater and pooled in a lake at the center, where the hippos wallowed. Over the millennia the crater floor had been covered with soil and pulverized pumice and lava, and was as verdant as anything in East Africa. From the crater rim the rudimentary road drops precipitously through acacia forests and into the grasslands covered with candle lilies, meadows of buttercups, blue hyacinth, marguerites, flame gladioli, and clover.

But it was the crater's violent origins that interested me.

From the summit of Ol Doinyo Lengai, or Mountain of God of the Maasai, you can clearly see Kilimanjaro and Meru, now the two largest mountains in Tanzania. It's a bruising eight-hour hike to the summit from the base camp below, but it's worth it, for you can also peer over the rim, cautiously, and stare into the interior of the volcano, which is still active; wisps of sulfurous gases and occasional bubbles of lava can be plainly seen. The hills are old, as old as the gneiss and granite outcroppings that mark the Serengeti plains, dating back maybe two hundred million years or so.

But some twenty-five million years ago, a very short period in geological time, dramatic changes came to the East African plains. The whole of East Africa began to crack away from the west and tore apart, causing massive gashes on the surface. Between the largest rifts, the land began slowly to subside and the Earth's crust softened and thinned. Magma bubbled to the surface, forming immense lava beds, then a series of volcanoes, which blew in turn, spewing vast quantities of lava and oily ash into the atmosphere and forever changing the landscape. Kilimanjaro, of course, and Meru, but also Lemagrut,

Sadiman, Oldeani, Olmoti (at 2.2 miles a shallow crater that is home to many grazing animals and at least one spectacular waterfall), Sirua, Lolmalasin, and Empakaai (which now holds a soda lake) and Ngorongoro thrust up along the Eyasi Rift, leaving behind huge cliffs at Lake Eyasi. Two much newer volcanoes formed along the Gregory Rift to the northeast of the Empakaai caldera. Ol Doinyo Lengai is one of these.

Ngorongoro, now extinct, was as tall as Kilimanjaro, then. The lava that filled it made a solid sort of lid, which collapsed when the molten rock subsided again, forming the caldera that we see today. Both Olmoti and Empakaai, smaller than Ngorongoro, collapsed the same way.

There wasn't just one cataclysm. The eruptions went on, and on, and on . . . new eruptions every million or so years. None were as great as the first, but great enough to cause massive clouds of ash and dust to settle on the landscape. As a byproduct, as it were, of these immense explosions and collapses, the East African forests turned into grasslands that not only supported the largest ungulate herds in the world but also changed the landscape within which a species of hominids lived, and (very probably) kick-started them onto a new evolutionary course. Some of this history can be read in the runes of the nearby Olduvai Gorge, where layers of sediment settled, trapping many of the creatures that lived there. The most recent creatures thus caught included our very earliest ancestors.

The Olduvai Gorge is some 260 feet deep and nearly 53 miles long, with two branches. It's not a very impressive place, neither picturesque like Ngorongoro nor as austerely beautiful as Serengeti. It is a rocky and barren landscape, but to paleontologists it is close to heaven, for in an accident of fate the gash in the Earth has exposed five levels of sediment from these earlier volcanic epochs, each rich in fossils. The volcanic ash has preserved in this unprepossessing little ravine an astonishing number of stone tools, hominid bones, ape and other animal fossils going back almost twenty-five million years, to the very beginning of the Great Rift itself.

Paleontologists love the fossils, but to the layperson the eeriest and most evocative of the discoveries are the tracks called "the Laetoli footprints," named after the branch of the gorge in which they were found. There, left in a muddy stream bank and subsequently covered with ash and fossilized, are imprinted the bare feet of our earliest ancestors. Three of them passed by that day so long ago, leaving two large sets of prints and a smaller one, presumably two adults and a child. They stride purposefully across the mud, surrounded by the prints of creatures much larger and more dangerous than they. These marvelous impressions have been carefully covered over again by curators to preserve them from the ravages of nature, but it is possible to superimpose in your mind the site, at the bottom of a broken, stony gorge, and castings of the footprints, and get a mental picture of the three little upright hominids as they made their careful way up the path (perhaps hand in hand, for the small tracks are parallel to the larger).

Here walked our ancestors, 3.6 million years ago, dreaming no doubt their wary dreams of warmth and wealth and well-being, bequeathing their dreams as well as their genes to the endless generations to follow. It is a humbling thought that all the world's saints and all the world's sinners, down over the endless millennia, can be seen in these tiny impressions—every warlord who was ever born, every mystic, every murderer, every poet, and every politician descended, somehow, from the people who left behind these small traces, and their kin. The three of them passed this way only once. But once was enough to make them immortal.

Perhaps they were a family, going home, making their way carefully through the many hazards that beset them. The ash that fell that night wasn't very thick, and they probably survived. I hope they survived. I'd like to imagine them there in the African night, listening to the rumbling of the Earth, pondering their fate and waiting for the dawn, like so many before them, and after.[1]

Everything here—volcanoes, plains, the Rift, ash strata, those fossils, these footprints, our own history—is therefore a byproduct of an Earth that was, quite simply, rent asunder, torn and twisted and upthrust, through processes that were (and are) at work far below the surface.

It is something we have come only recently to realize—how malleable, how very plastic, this seemingly solid Earth of ours really is, and how very flexible its apparently immutable surface. From our modern perspective, it is not hard to see why. If you want a homey analogy, picture a small rubber ball filled with some viscous liquid, rolling along the floor. The ball won't stay round. It will flatten along its axis and bulge along its "equator" as the inertial force of the spin pushes the interior in the direction of travel. That the Earth does this too was one of the earliest indicators that it is not solid at its core.

A recent series of passes by a German-American satellite has reinforced this notion of earthly plasticity and has discovered, too, some rather odder anomalies. Some parts of northern Canada, for example, have a lower gravity than the theories would account for. India, for its part, is in a (physical, not gravitational) "dip" or hollow of about a hundred meters—that is, it is 325 feet lower than it should be even on an ellipsoid Earth. Iceland, on the other hand, is elevated by 295 feet or so.[2]

The same satellite also showed that the Earth's surface is not at all static. It can change far faster than merely drifting continents, faster even than the seasons. On its multiple passes the satellite showed that the Earth's crust can dip a couple of inches under a heavy regional snowfall, or through changing water levels in the Amazon, or after torrential monsoon rains. Other research has shown that the deep tones of a thunderclap—not the crack that you hear but the almost inaudible infrasound rumbling that follows—can set off small seismic shocks that set the upper crust quivering.[3] Far from being the solid thing that common sense tells us it is, the Earth breathes, shivers, throbs, and pulsates in ways no one would have believed even a few decades ago.

That the moon causes tides in the oceans is a commonplace, of course. But that it causes tides in the apparently solid surface of the Earth is less well known, yet it does—the Earth's crust is "sucked upward" by the moon's gravitational pull twice a day, just as the sea is. Less than the sea, obviously—rock is not as malleable as water—but it can rise by a not unimpressive ten centimeters all the same.

The last ice age depressed those portions of the crust where the ice lay thickest. The Gulf of Bothnia in the Baltic is still rising thirty centimeters every twenty-eight years after the removal of this massive burden; Hudson Bay in northern Canada is rising even faster, whereas the part of the world I now live in, Nova Scotia, is sinking. Having "rebounded" quickly, it is settling back down again, so we're getting rising sea levels even without global warming.

Many people were astonished to learn that the Asian tsunami of Christmas 2004 was strong enough to actually change the Earth's rotation by a fraction of a second, setting off a flurry of apocalyptic blogs in the wilder shores of the Internet. But this is not so uncommon, after all. Changes of a few milliseconds are really quite routine. The seasonal distribution of water and ice, and even water vapor, can be enough. El Niño, in addition to causing droughts in the southern hemisphere and warmer winters in the American northeast, also changes the length of the day. Filling the Three Gorges Dam in China altered the rotational period enough to be measured by amateur astronomers.[4] If a dam below a 250 mile-long lake burst—Three Gorges is a good example— it would not only kill hundreds of thousands of people but also make the Earth itself wobble, and slow a little.

The Earth's yielding surface is one way to measure just how fluid the core is. Scientists can calculate the yielding that would occur were the Earth completely solid, and compare it with the yielding they actually measure, and thereby determine just how "soft" the center really is. This is the planet's "Love Number," not after any goddess of yearning but after the Oxford mathematician Augustus Love, who studied elasticity and wave propagation—"Love waves" are surface seismic waves that cause shifting of the Earth during an earthquake.

Love also figured out that a body with the composition of the Earth would actually vibrate if struck hard enough, and that the tone and period of the vibrations would depend on the character and composition of the interior. These vibrations—the Earth's ring tone, as it were—are called its "free oscillations," or, in a somewhat more opaque description, "normal modes." If the Earth were a solid ball of iron, Love asserted, its ring tone would have a period of about an hour— that is, if struck like a gong, it would ring in and out in an hour.

Love was doing this by blackboard and chalk. His assumptions were proven, however, by the seismologist Hugo Benioff, of the California Institute of Technology. Benioff was a gifted tinkerer and cobbled together many of the instruments still used in earthquake studies, such as a seismograph and a strain instrument that measures the stretching of the Earth's surface. In 1954 he detected a vibration period of fifty-seven minutes after the Kamchatka earthquake. His work was confirmed in 1961 by Bruce Bolt, another California seismologist (and creator of the California Academy of Science's simulation of the 1906 San Francisco earthquake that has drawn thousands of tourists since it was created in 1998). Bolt found a fifty-four-minute vibration after the 1960 Chilean earthquake and also tracked a very Benioff-like series of shorter-period vibrations that were harmonics or counterpoints of the greater ringing.

The size of the Earth has been known for more than two thousand years, although the number was lost or disputed when the Earth went from being round to being flat again in the Middle Ages (under the prodding of Christian heaven-watchers, the Greek philosophers' work was erased in Europe and preserved only by Muslim scientists in the Middle East and North Africa). The true size was first calculated by the Greek philosopher Eratosthenes, using Pythagorean geometry and extrapolating from a known distance, that between Syrene and Alexandria. His notion, that the circumference of the Earth was twenty four miles, is a mere fifty miles off the modern best estimates.

But, until very recently, no one knew what lay under the surface. It was sort of round, yes, but what was in the middle? Lost kingdoms? Solid rock all the way through? Glittering caverns of gold? Water? Chambers with great rushing rivers and mighty stalactites? Satan's minions? There were many guesses.

No one could yet go and see. The deepest cave systems went down only a half mile or so. Mines were no help—they were mostly shallower than the greatest natural caves. So it was left to theory, and later to the indirect measurement of seismic waves, to calculations of magnetic fields, and to some inspired guessing.

In 1906 Richard Dixon Oldham, then president of the British Royal Society, discovered a barrier when shock waves were deflected at a certain depth, and deduced that Earth had a core. It was already suspected that the Earth had been assembled from space debris—more precisely, that it grew from the slow accretion of solid particles drifting in from a large nebular cloud—and it became possible to suggest how, under the bonding influence of gravity, the young planet became ever more stratified. The heavier metals sank to the core; the heavier rocks, the basalts, out from that; then the lighter rocks, the granites. The very lightest materials, derived from comets, became the air and the oceans. After a hundred million years of gradual cooling, a sort of scum formed on the surface—that's the crust, the stuff we live on, and the stuff that shakes in earthquakes.

Thirty years later the Danish seismologist Inge Lehmann deduced that at the Earth's center were two concentric cores, a solid nickel and iron inner core of about 758 miles in radius (it may even be, bizarrely, one huge iron crystal), and a molten outer core, composed mostly of liquid iron, nickel, and sulfur with traces of other elements, with a radius of about 1,400 miles. Speculative theories suggest that the innermost part of this outer core may also contain heavier elements like gold, mercury, and uranium, the so-called trans-cesium elements (heavier than atomic weight 55). The two cores together are only about 15 percent of the mass of the planet. It's the molten outer core that generates Earth's magnetic field, in the same way a dynamo does. The inner core is too hot to hold a magnetic field; recent evidence suggests that the inner core may actually be spinning faster than the rest of the Earth, by maybe 0.3 to 0.5 degrees compared with the rotation of the surface.[5] Outside these double cores is the mantle, by far the largest bulk of the Earth, about 1,865 miles thick, making up perhaps 70 percent of the Earth's volume. The geology texts will tell you that "very little is known about the lower mantle," but this is something of an understatement. In fact, nothing at all is known about the lower mantle, except that it is seismically uniform, probably solid, and probably rock. The mantle as a whole is conveniently divided into 4 layers: the upper mantle (20 to 255 miles), the transition zone (255 to 416 miles), the lower mantle (416 to 1,739 miles), and D, the rest, for which they apparently couldn't come up with a good moniker. The upper mantle is made of a rock type known as periodite.

On top is the crust, a thin layer of solid material varying between 6 and 43 miles in depth. The continental crust is relatively thick (up to

40 miles or more), and made up of a great variety of rock types; the crust under the oceans is thinner (usually about 6 miles), and largely basalt. The crust is only about 1 percent of the total mass of the Earth; its temperature ranges from air temperature at the surface to somewhere around 1,650 degrees Fahrenheit near the upper mantle.

The crust and the top layer of the outer mantle are together called the lithosphere; the lithosphere, in turn, "floats" on a thin layer, the asthenosphere, a zone between 9 and 217 miles thick, made up of material that flows—sometimes upward through cracks and faults, rising to become volcanic islands or hot spots. The upper crust is made up of massive solid rocky plates that move independently of each other at a few inches a year but over geological time many thousands of miles. They usually move steadily and pile up, causing mountains, or tear apart, as in the Rift Valley. Sometimes, because surface rock is not elastic but brittle, the plates break along fault lines. We call those earthquakes. The explanation of why and how these plates move is called plate tectonics.

"The Earth," says Richard Fortey, "is like an avocado. The mantle is the edible flesh above the core. The crust might be the skin of the fruit. The motor of the Earth churns over in the mantle; it is where mountains are born and the plates die. It is the deep unconscious of our planet, the hidden body whose bidding the continents obey."[6]

"Rocks do not lie. They do, however, dissemble as to their meaning," Fortey said this too. What he meant was that it is not easy to see what rocks do, for they move at geologic time scales. And it is harder still to see why they do what they do, to discern patterns in the apparent chaos.

Still, that's the task of science.

The development of the theory of plate tectonics took a long time and much beavering away at the margins of theory. One of the barriers to understanding was the assumed age (or rather youth) of the Earth itself. For most of historical time, the Earth was assumed to be a fixed world, unchanging except at the whim of gods. For much of the Christian period, at least, the age of the Earth was determined not by observation but by Holy Writ; the genealogies of Genesis,

Interior of the Earth

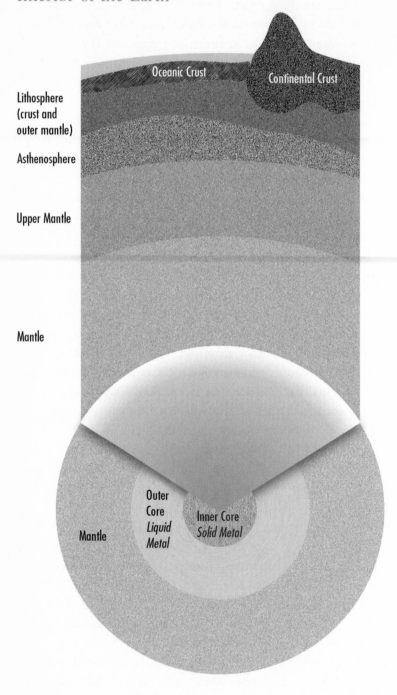

Oceanic Crust

Confinental Crust

**Lithosphere
(crust and
outer mantle)**

Asthenosphere

Upper Mantle

Mantle

Mantle

**Outer
Core**
*Liquid
Metal*

Inner Core
Solid Metal

counted backwards, suggested that Adam and Eve, who after all were there at the Beginning, were alive maybe four or five thousand years ago. Some modern Christian fundamentalists, stung by apparent proof that human societies go back further than this, have finessed the explanation—earlier generations of humans lived much longer, maybe nine times as long as modern humans (viz. Noah, who lived to nine hundred), and so . . .

This notion of a youthful Earth meant that change, if it happened at all, must have been sudden, or catastrophic, in nature. Noah's flood was the prime example.

This oddly provincial world-view didn't begin to change seriously until the nineteenth century. The first part of the theory to go was the reliance on the supernatural to explain mountains and oceans. Then the notion of sudden change itself came into disrepute. The evidence that was accumulating in the geological record suggested that if the known changes happened in the time allotted for the Earth's existence, then the changes must have been much greater, and much more catastrophic, than first thought. From there, it was but a short step to speculating that rather than relying on greater catastrophes, more time was needed. Perhaps the Earth was older, much older, than hitherto thought. It could be as much as, say, a million years old? Maybe even more?

It was left to a mathematician turned geologist, Charles Lyell, to at last provide a clear way of thinking about the state of the world. In the early years of the nineteenth century he substituted a rational system of investigation, based on meticulous observation, for the belief in a series of catastrophes. You could use the present to observe the past, he declared. What was happening now was happening then. Rain, erosion, slow climate change, earthquakes, volcanoes, wind, and other natural forces could, given a vast enough amount of time, re-order the world. In order to make Lyell's theories work, the Earth had to become very much older than even the first skeptics had thought. A million years wouldn't do at all. Many, many millions were needed. His theory came to be called Uniformitarianism.

His opening salvo was *The Geological Evidence of the Antiquity of Man*. But his most enduring work, the massive multi-volumed

Principles of Geology, published from 1830 to 1833, made his case that currently observable geological processes were adequate to explain geological history. Darwin had a copy of Lyell's magnum opus with him on the *Beagle*; it was the foundation of modern geology.

At the front of his book, he included an elegant engraving of the Temple of Serapis in Italy, later the subject of a monograph by a colleague, Charles Babbage, published in 1847, *Observations on the Temple of Serapis, at Pozzuoli, near Naples, with remarks on certain causes which may produce geological cycles of great extent.* Babbage, and Lyell before him, pointed to the fact that some of the surviving marble columns of the temple were 230 feet and more above sea level, but showed clear signs of being bored into by ocean mollusks, thereby proving that not only was the temple built above ground, it was then later submerged and re-emerged again, all of those in historical times. Push these processes much further back, Lyell speculated, and mountains can easily be made where only plains stood before.

He was right of course, but also wrong. Because he didn't yet understand what was going on *down there*.

The notion that the interior of the Earth was hot enough to actually cause convection currents was first advanced, as far as anyone knows, by an English vicar, Osmond Fisher, in the middle of the nineteenth century. No one paid him much mind. Dotty vicars were two a penny in Victorian England. That he was a first-rate scientist with a meticulous investigatory bent was for some while ignored. Still, in 1881 he published *The Physics of the Earth's Crust*, in which he suggested that Alexander von Humboldt's (and his) theory of moving continents could be explained by convection currents in the Earth's deep interior. This wasn't yet called plate tectonics, but it was close. "Continental Drift" would have to do for a while yet.

Humboldt (more properly Friedrich Wilhelm Heinrich Alexander Freiherr von Humboldt) was one of the first to propose that Africa and South America were once joined. Fisher and Humboldt were both ignored, or when they weren't ignored they were ridiculed. Lyell's steady state theory remained "true."

But change was in the air. Eduard Suess, a botanist who first pro-posed the notion of land bridges for the distribution of animals and plants, also suggested that continents could rise and sink. He couldn't figure out exactly how (except through the agency of some divine genie) but the evidence suggested that they might. It was left to a German meteorologist, Alfred Wegener, to pick up on Humboldt and suggest that the continents we see today were not always in the same place. In December 1910 he scribbled a note to his fiancée: "Doesn't the east coast of South America fit exactly against the west coast of Africa as if they had once been joined?" he asked. "This is an idea I'll have to pursue." Five years later he published his first approximation of the theory that came to be called Continental Drift, in a monograph titled *The Origin of Continents and Oceans*. In a subsequent edition he went further: "The geological evidence shows that about 300 million years ago, all of the continents were joined together. I will refer to this super-continent as Pangea, meaning all lands. It appears that Pangea began to break up about 200 million years ago." His super-continent Pangea was surrounded by a super-sea he named Panthalassia. His evi-dence included the distribution of fossils, the nature of shoreline rocks, evidence of glacial activity, and the relationship of mountain ranges on different continents.

Wegener didn't know how it happened, only that it did. Perhaps, he suggested, the movement could be caused by the Earth's spinning on its axis, through some version of centrifugal force?

It's fair to say that his ideas found a cool reception. The whole thing seemed so improbable. The president of the American Philosophical Society went on record to say that Wegener's theory was "utter damned rot." How could whole continents actually move? The thing was impossible. . . .

Still, someone did an English translation of Wegener's book in 1922, and then a French translation in 1936. The consensus was still that Wegener was a crackpot, but his evidence, on the other hand, seemed interesting. . . . And soon more "super-continents" were sug-gested—Gondwanaland, Tethys, and Laurasia among others.

A year after the French edition, the South African Alex du Toit published *Our Wandering Continents*, in which he summarized dozens of evidence strands and suggested an even more fluid planet than Wegener's: perhaps the super-continent suggested by Wegener had existed, but only in passing—Pangea had existed, but only as a point in time. Perhaps continents and land masses had been split apart and rejoined many times since the Earth was formed.

The next piece of the puzzle was provided by the physicist turned geologist Arthur Holmes, who had worked with radioactive isotopes in the 1920s and 1930s and knew how much heat radioactive decay could cause. Perhaps this heat could create convection currents in the deep substrata, solid yet sluggishly flowing, over millions of years? By the 1940s it was pretty much accepted that solid pieces of crust moved about on a more mobile inner layer, and that these "rafts" of rock are being folded and broken, lifted and lowered, in a continuous process of creation and destruction, all of it complicated by the tug of the moon, the rotation of the Earth, and the eccentricity in the Earth's orbit. The Canadian geologist Tuzo Wilson, in a paper published in *Nature* in 1965, was the first to call them plates. His theory of how it worked was published two years later.

But if chunks of continental rocks dived into the deeps, perhaps all the way to the mantle-core boundary, where was the new rock coming from to replace it?

The last breakthrough was the notion of the sea floor spreading. The expression was first used by Robert Dietz in a paper discussing the upper mantle convection theory, which spread apart the ocean floor in ridges and pushed the new rock under the continents through "trans-current faults." His paper was followed by another from Harry Hess, a Princeton geologist who served during the Second World War on a naval supply ship equipped with sonar; it was Hess who mapped the series of undersea ridges that girdle the Earth. It was soon discovered that the rocks, known as MORB for Mid Ocean Ridge Basalt, that were closest to the ridges were younger than those farther away, and that "new" rock was constantly oozing from the deeps. The theory of plate tectonics was finally in place.

Plate tectonics ("tectonics" either from the Greek meaning "without strength" or from the Greek meaning "to build," depending on your authority) is "the grand unified theory of geology," as an essay in *Nature* put it. "Everything we see today, from the abyssal plains of the oceans to the heights of the Himalayas, was shaped by plate tectonics. As far back as there has been complex life, and perhaps even before, continents have come together and moved apart in a dance that has altered climates and geographies, opening up new possibilities for life and sometimes closing down old ones."[7] In that sense, the world is one immense system, four and a half billion years old.

Literally dozens of theories exist on how plate tectonics work. But all insist on at least three conditions: the existence of rigid but mobile plates at the surface; the ability of those plates to move apart through sub-ocean spreading with its consequent spilling upward of new crust; and the requirement that plates must on occasion dive beneath each other at subduction zones.

So there it is: the Earth is everywhere shifting, moving, turning, seething. We live on the surface but are "moved and bound by deeper things," as Richard Fortey put it.[8] Everything is linked in ways we hadn't realized before—every patch of Earth is bound to every other patch, even to the point, in long geological time, of becoming those other patches; our current geography is just an accident of time, though our lives are too short to know it. We ourselves, and everything we live on—village, farm, city, island, continent—are being recycled, steadily, relentlessly, inevitably. Oceans are continually opening (the Atlantic is still getting wider) or closing (the Mediterranean is shrinking from the advance of the African landmass). The continents themselves are not only moving but being drawn down into the molten mantle, there to be recycled and eventually lifted back to the surface.

No one knows, really, exactly how many plates there are. The number changes all the time, as new faults are discovered, new theories proposed. Some have counted eleven "major" plates and maybe twenty smaller pieces. Others say there are between six and thirty-six major plates. The majors usually have obvious names: African Plate, North America Plate (still being born undersea, and in Iceland), Eurasia Plate, Pacific Plate. Others have more obscure names: the Nazca Plate, the Resurrection Plate, the Juan de Fuca Plate. In some places where plates collide, both are too light for one to dive (or "subduct") under the other, and crumpling occurs. This is true, for instance, in the Zagros Mountains of southern Iran, where the Arabian Plate is impacting the Iranian plate. Other places, such as Nevada, are complex and contain "extensional" plate activity (two plates pulling away from each other) as well as "transform motions," which occurs where two plates slide along each other. And sometimes plate fragments spin, as when a mass, such as an island or an undersea ridge, blocks the downward flow of the plate. Tokyo is sitting on a zone like this. The collision causes fragments, then the fragments are set spinning by the torque that the lower plate exerts as the unblocked regions continue to subduct. The mechanism may also trigger the stretching of the upper plate that sometimes causes it to split.[9]

The ponderous progress of the plates, by about an inch a year (about as fast as fingernails grow), with the consequent sea-floor splitting and formation of new ocean and crust, followed by the deep diving of old crust into the molten depths, is now known as the Wilson Cycle, after Tuzo Wilson. Many cycles can be found in the natural world—the slow cycling of the deep oceans comes to mind—but this is the slowest, grandest, and most profound of all. It can take maybe two hundred million years or so to complete.[10]

This slow process can mask events of extraordinary violence. This is a dynamic planet, with at least 1,500 volcanoes (660 of them active at this time), 44,000 earthquakes so far cataloged, and some 170 impact craters, evidence of extraterrestrial invaders of awesome potency. In an average year some 60 volcanoes go off and more than 160 earthquakes of magnitude 6.0 or more, enough to cause multiple casualties and substantial damage.

The most violent convulsions take place in the "convergence zones," where places collide or tear apart. The Eldgja Rift in Iceland is one of these. So is the San Andreas Fault, perhaps the best known, and one with a massive potential for calamity. Another is the Great Rift of East Africa.

The day after I had sat with Viola at the summit of the Ngorongoro Crater, I went down the *en tiak* itself, the crater wall, with one of his

Significant Tectonic Plates

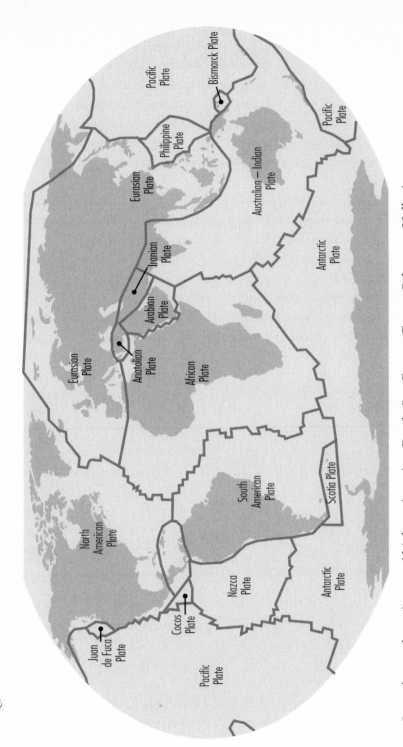

Some lesser plates: Aegean, Altiplano, Amurian, Banda Sea, Burma, Easter, Galapagos, Hellenic, etc.

compatriots. We saw, as promised, the usual sights—wallowing hippos; giraffes with their curiously feminine eyelashes; a python as thick as a thigh; baboons; a cheetah sitting on a termite mound for a better view; many other four-by-fours, each plump with tourists, cameras clamped to eyes; and wildebeest, with their patented look of cross-eyed idiocy. And a lion that snarled as we went by, lifting her muzzle from the ribcage of a little gazelle, a life and death scene—I thought then—as old as time.

But it wasn't as old as time. Life has been around only for a heartbeat by the standards of the rocks of this place. The cataclysm that created the crater would have wiped out not just a pretty gazelle but any life that existed within about 600 miles. In the hierarchy of violence, and by the standards of the cataclysmic upheavals of the past, predator–prey life-and-death cycles are puny and insignificant. The lion was triumphant, but in a blink it and its kind will be gone forever, and the planet won't care.

FIVE

Our Ever-Changing Climate: Ice Ages Now and Then

The long drifts of weather we call climate…
—Frank Herbert, ecologist and author of *Dune*

The most charged debate of our times, at least to those of us not living in a war or calamity zone, is about global warming and consequent climate change. Every week, it seems, there is a new book, or a new Oscar-winning film, or a new documentary or news film, about the dire things we are doing to the Earth's climate. Just on my own little bookshelf sit *Field Notes From a Catastrophe*, *Heat*, *The Weather Makers*, *An Inconvenient Truth*, and a dozen others—what a catalog of gloom! What a list of awfulness! And all the more dire for being all too plausible.

So much discussion, then, about climate change—but what is this thing we are changing? The inbuilt assumption of the whole debate, the entire rationale for the cries of alarm, is that we are changing things for the worse—that is, current climate good, climate change bad. But without examining the merits of the case for or the case against—let's just stipulate for the moment that climate change is something we should not be doing, but are—still, the former assumption goes largely unchallenged. In fact, *is* climate good? What climate? Is the current climate optimal? What, indeed, is climate? How stable is it? What, indeed, makes it "good"? And what, if anything, can be done about it? Even if we reverse our course completely and soon, climate will change anyway. What then?

Is climate good because we have adapted to its current iteration? Is it optimal for, say, equatorial Americans and Africans, sweltering there in the tropics? Is it good for the Inuit, those most adaptable of humans, clinging to a precarious existence on the ice? Is it good for the farmers scrabbling a livelihood on the edges of the world's great deserts? Is it good for the wretched living on the floodplains of Bangladesh? Why is so much of our one and only home in fact uninhabitable in the current climate?

The other great untested assumption in this ongoing debate is the idea of biodiversity.

Biodiversity is a good thing—we simply accept it as a truism. But how true is it, really, and from what perspective? The great hothouses of diversity are the tropics, and our view of diversity includes all the beautiful and complex creatures we have come to know so much about—as well as other creatures. Is every virulent parasite, every burrowing worm, every poisonous frog, really a good thing? Must we clasp to our bosoms the bacteria that produce smallpox, the Ebola virus, leprosy? In this context the earth scientist James Lovelock has said, it seems to me, something truly revolutionary. He has suggested that to the planet, or to Gaia as he has personified the forces that keep Earth's climate in some kind of equilibrium, biodiversity (love you, pandas!) is either irrelevant or a sign of disequilibrium, a consequence of instability, a sign of a planet in fever. As he put it,

> The biodiversity of the Amazon rain forest may be the Earth's response to the heat of the present interglacial, rather like a flushed and sweaty skin is our response to overheating. We have become so concerned over the fate of the rare tree and about rare and beautiful animals, we have become so excited about these collectibles because we have lost sight of the forest itself. Gaia's automatic response to

adverse change is driven by the changes in the whole ecosystem, not by the presence or absence of rare species.[1]

Most readers of Lovelock seem to assume he is one of the proto-typical tree huggers, against anything that changes the Earth, and for the sanctity of everything—forgetting that his Gaia is impersonal, and that to her (Lovelock has personified "her" to that extent) the ice ages so deadly to us are just another planetary regulatory device. Gaia's "goals are not set points, but adjustable for whatever is the current environment and adaptable to whatever form of life it carries." Evolution is fecund partly because it needs to be, and it needs to be partly because everything about the Earth is unstable. Rapid evolution, explosive mutation, guards against extinction, which is a careless byproduct of inevitable change. Gaia would be happy enough if grass were the only viable life form, with maybe something to crop on it (thus completing the carbon dioxide–oxygen cycle). We as a species are not really necessary, and nor are millions of others.

When it comes right down to it then, biodiversity is good only because of the inherent instability of climate.

If we know anything at all about climate, we know that it is changing anyway, regardless of our actions or inactions. Climate is what they call a dynamical system: constantly changing parameters, time scales, and patterns. The Sahara has not always been dry, the poles always cold, the tropics always torrid. We know there were ferns and jungle in Nevada, once and several times; the ice sheets covered most of France and North America down to Pennsylvania and beyond—my own house was a kilometer beneath the ice less than an eon ago, under the crushing weight of the glaciers. The oceans have not always circulated as they do now, regulating weather and bringing temperate air to northern regions; indeed, on a longer scale even the oceans have not always been where they are, nor the land masses around which they now circulate.

Our climate memories are not even good in the short term: because America's settlers found the prairie grasslands verdant, they assumed that was the natural state of affairs. Accumulating evidence suggests that they were fooled—drought and desert is the norm, and they (and we) have been living in a fool's paradise of very short duration, now over.

We have taken many of these short-term aberrations as the natural state of things. As Donald Kennedy, then editor of the journal *Science*, put it in a recent editorial:

> Nothing in the record suggests that an equilibrium climate model, such as the one of the last 10,000 years or so, is the right mode for understanding the future. We are in a highly kinetic system and in the past dramatic climate changes have taken place in only a few decades. Our comfort levels in the present Holocene may have heightened our sense of security, but the expectation that climate change is unlikely is not a reasonable position.

And that's without us doing anything at all.

We've been living in a little bubble of stability in a great sea of chaotic change. And we thought it was forever.

The climate is maintained, within this little bubble of stability, by an intricate series of cycles and forces, beginning with that distant nuclear reactor we call the sun, the source of all energy.

Deep inside the seething cauldron of the sun, thousands of miles beneath the corona, or what passes for its surface, is a constant cascading series of hydrogen fusion reactions. Every nanosecond millions of hydrogen atoms crash together, and for every four that destroy themselves in this furious suicide, one helium atom is created. Since four hydrogen atoms weigh slightly more than one helium atom, a fraction of the mass is lost each time, and this shortfall—*pace* Einstein—is released as pure energy. So much energy, indeed, that the temperature of the sun is maintained at a fairly steady 59 million degrees Fahrenheit. Some of that energy is radiated into space. A tiny fraction, about two-billionths, reaches the Earth. This might not seem a lot, but the sun is so massive relative to the Earth that the solar radiation reaching us is around 175 trillion kilowatt hours of energy every hour.

The sun has been doing this for maybe five billion years. In another five or six billion, it will burn itself out. It will go dark, or collapse into a red dwarf—or go nova.

So—five billion years to go. That's not bad.

But after knowing what the sun does for us directly, it gets more complicated.

The Earth revolves around the sun—it orbits the sun—every schoolchild knows this. But it doesn't orbit with quite the neat circularity that Galileo had depicted—the Earth and the other planets moving in beautiful circles around the sun, so many druids dancing to its life-sustaining flame. Nor is the world quite as round as the ancients thought, as the previous chapter suggested.

It was Isaac Newton who described the true shape of the Earth. He did this by pure inference, by calculation and not by observation, but he saw that the Earth must be oblate, not an exact sphere but bulging a little at the equator because of its rotation. He also saw that the Earth's orbit is not a circle but a leisurely ellipse, and moreover that this ellipse has an inbuilt eccentricity, a periodic elongation of the ellipse itself, for reasons still unknown.

He also pondered another and more mysterious motion, one that had been identified millennia earlier by the Greek astronomer Hipparchus of Nicea, who called it "the precession of the equinoxes." Like a spinning top slightly off balance, the Earth changes the orientation of its axis against the backdrop of the stars, by about a degree every seventy-two years, or a full circle in somewhere around twenty-six thousand years. Newton (having "invented" gravity in the first place, a notion much derided by his contemporaries) calculated that the precession would be caused by the complex result of the gravitational pull of the sun and the moon on the Earth's bulging waistline. The Pole Star to which the Earth's axis now points wasn't the pole star in Hipparchus's time, and won't be again in a thousand years or so.

Newton also described the third of the Earth's known changing cycles, called obliquity, referring to the tilt of the Earth's axis relative to the plane of the ecliptic (the ecliptic is the apparent path of the sun's motion as seen from Earth); it varies between 22.1 degrees and 24.5 degrees, the higher angle of tilt increasing the seasonal contrast most effectively at high latitudes—that is, winters will be colder and summers hotter as the angle increases.

All these events are cyclical, which means that they occur with predictable regularity. But the cycles overlap, sometimes reinforcing each other and sometimes canceling each other, and the actual effects on climate are hard to calculate. Orbital eccentricity occurs at two cycles, with periods of 100,000 and 400,000 years, on which another 40,000-year cycle is imposed. Obliquity recycles in 41,000 years, and precession at two cycles, with periods of 19,000 and 23,000 years. The amount of radiation—dubbed insolation—the Earth receives is therefore quite difficult to calculate, and even today calculations emerging from the periodicity of cycles are far from settled.

Still, it has become increasingly apparent that over the last sixty-five million years or so—at least since the dinosaurs perished—the Earth has experienced pretty constant and consistent change, drifting from one extreme (massive continental ice sheets) to the other (hot and humid greenhouse effects, even at the poles). The orbital rhythms are complicated by Earth's internal dynamics, including changing continental topography, ocean circulation and other factors, all of these changing constantly, albeit slowly. Some of the more dramatic changes in the period include the opening of the mid-Atlantic rift, the opening and widening of the ocean around Antarctica, the uplift of Panama and the closing of the seaway there, and, most dramatically, the collision of India with Asia. Each of these "internal" Earth events radically altered the planet's climate, sometimes dramatically and suddenly.

Another complication is that the sun itself exhibits cycles, some of them worryingly large.

The best known of these is the eleven-year sunspot cycle. That these "spots"—enormous flares in the sun's corona—are still called spots is an indication that we don't know very much about them. It is thought they are twistings of the magnetic field at the surface, though why this should be so, and why it occurs every eleven years or so, are not known. The sun's magnetic field generates a flow of electrically charged particles, the solar wind; every eleven years huge solar flares explode above the new spots, causing turbulence in the solar wind and sending fresh streams of charged particles into space. The last active phase was in 2003. In October that year astrophysicists spotted two spots, each the size of the Earth, each pouring out billions of tons of charged particles. Half of NASA's satellites malfunctioned as a result.

In November, as the second spot was about to dissolve, it let loose a tremendous explosive gasp. As Stuart Clark put it in the journal *Nature*, "By sheer luck it exploded into deep space and only caught us in the side wash. . . ." Mausumi Dikpati and her colleagues at the National Center for Atmospheric Research have developed a computer simulation that mixes the sun's internal magnetic dynamo with theories about how plasma circulates near the surface. "[In the next cycle] we expect between 30 per cent and 50 per cent more sunspots and solar activity than the cycle just ended," she said.[2] Sunspots have a direct effect on climate: during solar minima (low point in the cycle) Earth was 3 percent cloudier than at solar maxima (high point). Tantalizing hints suggest that sunspot activity can affect the influx of cosmic rays coming from deep space, which in turn can help seed cloud formation. If this is true, the Earth should be cooling off somewhat in the years 2012 to 2017.

Some scientists think it will do more than that and think they have detected another cycle, which will produce a major solar output decline between 2035 and 2045, dramatically cooling the Earth. The leading proponent of this theory is a Russian astronomer, Khabibullo Abdusamatov, who toils at the Pulkovo Astronomic Observatory in St. Petersburg. The most recent such cool-down period in the northern hemisphere was between 1645 and 1705, he pointed out. The resulting period, known as Europe's Little Ice Age, left canals in the Netherlands frozen solid and forced people in Greenland to abandon their houses to glaciers. "Dramatic changes in the Earth's surface temperatures are an ordinary phenomenon, not an anomaly," he said, "and result from variations in the sun's energy output and ultraviolet radiation."[3] Needless to say, his theories are hotly disputed, and in any case even if true might be outweighed by anthropogenic, or human-caused, warming.

Robert Ehrlich, an American astrophysicist at George Mason University in Virginia, is another who has suggested that periodic solar variability has been the cause of global temperature cycles, with periods ranging from the eleven-year sunspot cycle to cycles as long as two thousand years; he believes that tracking what he called resonant thermal diffusion waves from the sun can explain many details of the temperature records for the last 5.2 million years. In effect, he suggests that there is some kind of a dimmer switch deep inside the sun that cycles on and off at periods varying from 41,000 to 100,000 years, thereby making it neatly coincide with known ice ages.[4]

As though that were not enough, it seems there might be factors beyond the sun that also impinge on climate, albeit often in very long cycles. A report in *Physical Review Letters* in 2002 reproduced research that seemed to show that ice age epochs on Earth are historically correlated with the solar system passing through the spiral arms of the Milky Way.

These spiral arms are not permanent fixtures, but rather are transient effects from density ripples traveling around the galaxy. Nir Shaviv, of the University of Toronto and Jerusalem's Hebrew University, has deduced Earth's exposure to cosmic rays by considering the cosmic ray exposure of forty-two iron meteorites, and found that the cosmic ray flux varies with a period of about 143 million years, which correlates well with both the geological records of ice age epochs and the solar system's location relative to the spiral arms. "Our current position in the minor Orion spiral arm should lead to cosmic ray fluxes about half of what we would receive in a major spiral arm," he says. His model places us in the wake of a major ice age epoch and is consistent with the global temperatures that we are now experiencing.[5]

Despite all this knowledge and these theories, paleoclimatologists are still faced with events they are at a loss to understand. They call them, logically enough, aberrations—defined as anomalies that stand out well above normal background variability.

The three largest known anomalies all occurred at epoch boundaries, as understood by geologists, and all may have had a profound effect on life (as we shall see in Chapter 6, on extinctions). The most prominent was the Late Paleocene Thermal Maximum, which occurred some fifty-five million years ago with a sudden spike in what we would now call greenhouse effects—rising sea temperatures, even in the deep oceans, and much higher humidity and rainfall. It took almost two hundred thousand years to recover, for the planet to be restored to some kind of thermal equilibrium. This anomaly may have been triggered by a catastrophic release of undersea methane (see Chapter 10).

The other two aberrations were both extreme cold episodes, one at 34 million years ago that started with the startlingly sudden appearance of massive ice sheets in Antarctica and led to a 400,000-year glacial period; the second was at the Oligocene/Miocene boundary, some 23 million years ago, which consisted of a brief but very deep glacial maximum, lasting some 200,000 years. (At the same time, for reasons not yet understood, atmospheric carbon dioxide content fell dramatically at these points.) Both these aberrations may be related to orbital fluctuations, but no one is yet sure. Still, "their mere existence points toward the potential for highly nonlinear responses to [climate changes] forcing in climate, or the possibility of unexpected anomalies."[6]

So what causes the ice ages, the longest climatic cycle we know? Why is the Earth periodically plunged into the cosmic deep freeze? For millions of years, after all, the Earth has undergone a recurring cycle of ice ages, each typically lasting about a hundred thousand years, but many lasting much longer. The ice ages, it could be argued, are the norm, merely punctuated by shorter periods called interglacials. We're in one of those now. Why does it happen—and how does the Earth recover?

It is now almost universally accepted that the Earth's orbital cycles are the prime cause. The theory was first proposed in 1842 by the French mathematician Joseph Alphonse Adhemar, and then again by James Croll, a self-educated Scottish physicist, in 1875, but it wasn't until the 1930s that it was widely believed, after calculations were published by a Serbian mathematician and astronomer, Milutin Milankovitch, who worked at the University of Belgrade. The first iteration of Milankovitch's theory proposed regularly recurring ice ages based on orbital precession—that is, there should be recurring cold and hot cycles on a twenty-six-thousand-year periodicity. Currently, the Earth's closest annual pass to the sun is the first week of January, when the northern hemisphere is pointing away from the sun; this means that the north gets its least sun just when the Earth gets the most. You would expect, then, that the southern hemisphere would be extra hot, and it is, but the effect is somewhat masked by the fact that most of the southern hemisphere is sea, which moderates the heating. Some thirteen thousand years from now, Milankovitch argued, the northern hemisphere's midsummer would occur when it points at the sun; this would mean more extreme seasons. Northern winters would be much colder, causing more snow to fall, increasing the Earth's reflectivity (called its "albedo"), thereby cooling it even more dramatically.

Since ice ages don't in fact occur in twenty-six-thousand-year cycles, Milankovitch and others revised his theory to include the cycles of eccentricity and inclination; recent studies of oxygen isotope content in the deep ocean have been in full agreement with this Milankovitch-plus.[7]

One of the scientists who argued for the acceptance of the Milankovitch theory was George Kukla, now a micropaleontologist at the Lamont-Doherty Earth Observatory of Columbia University, in New York. Kukla, more controversially, was one of the many scientists who argued in the 1970s that the Earth was about to enter a new ice age—a prediction that seemed to be borne out by temperature studies and that led to a famous *Newsweek* magazine cover story in 1975 titled "The Cooling World." Although it is now suggested that the cooling trend, while real, was a short-term blip in a longer warming trend and was caused by massive infusions into the atmosphere of industrial pollutants that screened out sunlight, since banned and reduced, Kukla and some others still believe they were right in 1972. The current warming trend, he has said, is normal. Ice ages are always preceded by warming trends.[8]

When the Earth is in the grip of an ice age, when the whole planet is in the deep freeze, and massive ice sheets are pressing down on the continental landmasses, how does it recover? What makes it warmer again?

The simple answer is that it eventually passes into another phase of the cycle that caused the freezing in the first place. But considerable feedback effects are also thought to operate that can make the recovery much faster. One is this: the freezing lowers the ocean levels by a considerable margin, thereby relieving pressure on the stored banks of concentrated methane hydrate in the ocean depths, which are then explosively released, causing a greenhouse effect, and reversing the cold. Just such a methane release might have triggered the Late Paleocene Thermal Maximum, as suggested above.

It is climatic balancing acts like this that led James Lovelock to propose his Gaia theory of planetary regulation. It can look astonishingly purposeful.

Climate, then, is a dynamic system, that is constantly changing—from a layperson's point of view almost capriciously. It has changed in the past without our help, and will change again, sometimes slowly, sometimes precipitately. For the last ten thousand years or so, a very short period in the geological scale but one that encompasses all the history of human urban civilization, climate has been in a state of equilibrium—not stable, exactly, but balanced on a knife edge. Like all balances, it is easily disturbed, easily tipped into another state, into abrupt change—and here I mean abrupt on a human scale, not a geologic one: years instead of decades or even centuries. What these tipping points are we can only guess. It may be a sudden flaring of the sun, a hole in our magnetic shield that protects us from cosmic radiation, a catastrophic event of the sort that this book explores, like a sustained volcanic eruption or the sudden release of methane into the atmosphere. Or, indeed, a change in the accumulated balance of greenhouse gases in the atmosphere.

But should we then just give up or panic? Change, almost any change, may make a few parts of the Earth more benign than they are now (Siberia as the breadbasket of the planet, Nova Scotia as the new Burgundy?), and it may not, in the end, be catastrophic. Some deserts will surely expand, but others, just as surely, will get wetter. The poles will surely be more benign places to live, parts of the mid-latitude subtropics will surely get worse. On the whole, the evidence that the planet as a whole will be worse off is very thin. Some parts yes, some parts no.

Whatever its effects on the Earth itself, climate change will certainly be wrenching for our species, because we have evolved a way of living in the comfort of the Holocene Stability. It will be incrementally bad for the poorer citizens of our planet, and incidentally worse for many of the other species that share the world with us. Our food production is based on the current climate, and thus too our lives. Our cities and towns are where they are because the coasts and rivers are where we find them—a consequence of the current climate—and rising sea levels mean drowning cities and, at best, millions of environmental refugees.

Change will happen, whether through natural causes or through our own actions. It is probably already too late to head it off. Which leaves us with two points to ponder. First: climate change needn't be Armageddon—we will almost certainly be able to adjust, though at great cost. And second, why contribute to its speedy arrival? Why bring on a crisis whose onset could otherwise be delayed? Why accelerate calamity? The self-correcting systems that James Lovelock calls Gaia may afterwards restore some sort of equilibrium. Better we should give ourselves as much time as possible to adapt to the new reality.

SIX

Fragile Life:
The Conundrum of
Mass Extinctions

More than any time in history mankind faces a crossroads. One path leads to despair and utter hopelessness, the other to total extinction. Let us pray that we have the wisdom to choose correctly.
—Woody Allen

L ife is a miracle, yes, but its continued existence is far from a sure thing. God does not play dice with the universe, the unbeliever Albert Einstein famously said in quite another context, but on the evidence, it seems he does. If you look at the paleontological record, God seems as reckless with life as a binge gambler convinced his losing streak is over. Life was very nearly extinguished on Earth not once, but at least five times.

I 've crossed the Karoo in South Africa more often than I can count. When I was a kid, it seemed excruciatingly boring—flat, barren, treeless, featureless, ocher and dun in color, apparently endless. In subsequent years I learned, under the coaching of my art director wife, to better appreciate this gently sloping South African inland basin, and to appreciate the subtle color palette, the duns and browns of the endless grasslands infused with delicate violets and pinks. I even learned to like the pacing, and not to mind the hundred or so miles between gas pumps, usually located outside the *algemene handelaar*, or general store, the only businesses in the sleepy little farming communities through which we sped—De Aar, Laingsburg, Beaufort West, Three Sisters, their main streets one hundred sixty-four feet wide, enough for the turning of a whole span of oxen.

The whole Karoo basin is high up, over a three thousand feet in elevation and higher. If you drive from Cape Town on the coast you climb Bain's Kloof Pass, going up sharply from the coastal plains, rising six thousand feet or so up a winding, twisting mountain pass, then you reach the summit . . . and stay there, or higher. You don't go down again when you reach the summit; you'd stay at that altitude for another six hundred miles or so, or three thousand more if you were traveling "up Africa."

This Bain after whom the pass was named was Andrew Geddes Bain, an engineer who made his living building roads, punching several passes through the coastal mountains, making deep cuts in the escarpment as he did so. Bain had become enthusiastic about geology after reading Charles Lyell's *Principles of Geology* in the 1830s, and his amateur geologist's eye kept finding fossils buried in the rocks his road crew were blasting away. He pried some of them out. They looked like reptiles, but sometimes rather dog-like reptiles, and he wasn't sure what they were. Eventually, he shipped some of them to London, to an acquaintance named Henry de la Beche, a member of the Geological Society of London.

Those fossils were a key to unlocking the Great Dying, the mass extinctions at the end of the Permian Era, some 251 million years ago, the greatest die-off in the known history of planetary life and "the most fascinating mass-murder in Earth's history."[1] Almost everything died. It was a near thing.

The Karoo wasn't arid then. It wasn't even in South Africa of course, since not only was this before any politics, it was before Africa had separated from the ur-continent, Gondwana. Geographically, it was closer to the south pole, but the climate of the time was not cold but warm and fertile, fecund with life, with plants, insects (including massive dragonflies of an iridescent blue) and plenty of the peculiar reptiles that had caught Andrew Bain's attention. Only sheep graze there now, but the rocks have trapped evidence of an ecology as diverse as that inhabiting the modern Kruger National Park. You can still see its denizens peeping from rock strata or, if you're not a hiker, in the exhibition rooms of the South African Museum in Cape Town. These were synapsids, mammal-like reptiles or lizard-like mammals, it's hard to say from looking at them. Vertebrates, in any case—Earth's first great alpha species, the first dynasty, the dinosaurs' predecessors, a dynasty that lasted an astonishing sixty million years. All gone now, gone in the Great Dying at the Permian's end.

The synapsids were a varied species. There were the beaked dicyno-donts, herbivores that browsed along rivers, smaller grazers like the piggish *Diictodon*, alpha predators like the gorgonopsians—carnivores with teeth like razors. Most of the Karoo specimens were dicynodonts, which means two-dog-tooth, so called because they typically had a short skull with a horn-covered turtle jaw and only two teeth—sharp canines. They were massive-bodied creatures, with stumpy legs and a rudimentary and rather comical tail. Synapsids had whiskers and body hair, and their most famous member was the dimentrodon, often mistaken for a dinosaur. This odd creature had a smiley face, sharp teeth, and what looks for all the world like a huge sail on its back.

At deep levels, where the oldest deposits are, many such creatures are found. The shallower the sedimentary rock deposits excavated, the fewer they become. By the most recent levels, all that is left is an animal known as *Lystrosaurus*, which looks rather like a piggy mastiff with overlong teeth. At the very top, dozens of *Lystrosauri* can be chipped from the rocks—so many that paleontologists have come to pretty much ignore them.

Lystrosaurus survived, no one knows why. Almost everything else died. An astonishing 96 percent of the species that lived in the seas died, including corals, bryozoans, arthropods, and many others. Almost three-quarters of everything that lived on the land died. Life itself nearly died (in fact there have been several books written about the extinctions that make this very point—see *When Life Nearly Died* by

Michael Benton, and *Extinction: How Life on Earth Nearly Ended 250 Million Years Ago*, by Douglas Erwin).

It didn't though. In the end, life triumphantly survived. The Great Dying opened the way first for the archosaurs, and then, after another bout of extinctions, or perhaps even before it, the dinosaurs. Early dinosaurs seem to have been small, bipedal—and fast.

The extinctions seem, from the evidence, to have been dramatically sudden. Dated radiometrically from intercalated volcanic ashes, they apparently took place over a mere few hundred thousand years and quite possibly much less, a catastrophically short period by geological standards.

So what calamity caused it?

The search for the cause, or causes, exposes a whole dismal range of possible horrors.

The prime candidate is volcanism, and the prime suspect a massive lava flow in Siberia, under the city of Norilsk. The eruptions that caused the Siberian Traps, so-called, must have been gigantic and prolonged—the lava that vomited out of the Earth now covers 15 million square miles to a depth of 900 or 1,300 feet—an area the size of Western Europe, with enough lava to cover every landmass on Earth in 20 feet of molten rock.

Basalt, which is what the Siberian Traps are made of, doesn't always come out of the classic conical volcanoes in rapid and explosive eruptions. Sometimes, as in Siberia, it simply pulses out of the ground, as though the Earth were bleeding glowing magma. There are numerous examples of where this kind of thing happens. The Deccan Traps in India, which date from much later, about sixty-five million years ago, were similar in type and scope. More modern examples are easy to find. The 1783 Laki eruption in Iceland was similar, though much smaller, and in 2006 hot mud started spurting from a hole in a field outside the Indonesian town of Sidoarja. It started in a rice paddy but soon destroyed the town, turning into a black mud lake stretching for many miles higher than the treetops, barely exposing rooftops and streetlights. It started oozing at 650 cubic yards a day; by February 2007 it had reached 170,000 a day and was showing no signs of slowing. It could go on for years, even decades. (In a nice Orwellian touch, the Coordinating Minister for People's Welfare, and one of the country's richest men, owns part of a drilling company that was blamed for punching a hole into the wrong place.)[2]

Laki is an interesting example. It wasn't a violent eruption, but it did pulse massive amounts of noxious gases into the atmosphere and lowered the temperature of the globe by nearly 2 degrees for a year or so. How much more acid rain could the much greater Siberian Traps have produced? How much more particulate matter thrown into the atmosphere, cooling the planet enough to set off a bout of glaciation, thus lowering the sea levels? Lowered sea levels, in turn, could have released the sea's methane deposits and carbon dioxide deposits, which, combined with the extra carbon dioxide from the eruptions, and even vaster quantities of carbon dioxide from the megatons of decaying animals and vegetation, could have reversed the glaciation and set off a runaway greenhouse effect. (We know most of the vegetation died, partly because a massive spike in fungi has been discovered in the right period, most of it a wood-decaying fungus whose population would have exploded with all that lovely dead stuff to feed on.)

In this scenario, nothing from space came to get the creatures of the Permian. This was a purely homegrown catastrophe.

There was, in all likelihood, a cascading series of disasters. It would have started when Siberia erupted. What caused the eruptions is not known. It could have been caused by what is now Norilsk passing over a planetary hot spot, a crustal weakness. It could have been the crustal plates of Pangea jostling for position as the continent prepared to tear itself apart. For whatever reasons, vents opened in a series of volcanoes, and lava began to pour into the open. As it did so, huge amounts of ash, dust, and debris were thrown into the air, some of them rising into the stratosphere to catch the jet streams, surrounding the Earth in days. The debris cloud would have stayed with the planet for months, years, even decades as the eruptions continued—there is some indication that the eruptions lasted a million years or more, if only intermittently. The first effect would have been to cool the Earth off dramatically, causing glaciation, the shrinking of the oceans, and massive die-offs of plant life and those species dependent on them. Shallow sea dwellers would have died out.

As the heavier particles fell to Earth, the planet would have begun to warm up. An early consequence would have been a much reduced ozone layer, and in fact mutant pollen and spores in sediments from the time hint at a sharp increase in ultraviolet levels. Increasing concentrations of methane and carbon dioxide, much of it produced in the eruptions themselves, would have overloaded the feedback regulatory systems. The ocean would have been oversubscribed with carbon (very likely fizzing like an opened champagne bottle), and the temperatures would have increased dramatically. Intensely acid rain would follow.

We've seen the effect of acid rain in our own times. Some years ago, when my wife and I owned a small farm in Ontario, the maples in our forest began to turn anemic and sickly, the direct result of an attenuated acid rain caused by coal-fired generating stations in Michigan and Illinois. It was minor, as it turns out: the generators installed scrubbers and our forest improved. But I've also seen the results of more severe acid rain—for example in northern Czechoslovakia, as it was then, in Bratislava province, where unregulated industry under the last of the communist regimes poisoned the air and ruined the countryside, killing all the forests and turning the Earth black. Even worse was in the Middle Volga of Russia itself, where chemical factories (ironically some of them making plastic "wooden" furniture) quite literally killed the countryside for hundreds of miles. The Siberian Traps eruptions, however, wouldn't just have killed vegetation for hundreds of kilometers. Their effects would have been felt everywhere.

The dying vegetation caused that fungal spike and another sharp increase in CO_2.

At the same time, atmospheric oxygen levels seem to have dropped, down from nearly 30 percent in the Carboniferous period to somewhere around 13 percent (current level is around 21 percent). Inefficient breathers would have quickly died out.

In the sea, things would have been just as bad. In the mid-1990s a team of British scientists from the University of Leeds found evidence of anoxia, an absence of oxygen, in rocks from the late Permian. They speculated that changing ocean currents—or even a complete cessation of ocean currents as Pangea broke up—abolished the circulation that refreshed the sea with oxygen. If this were indeed so, it would explain why most of the remaining marine life perished—the animals simply suffocated.

It's also possible, though not certain, that the changing seas would once again have disturbed what methane hydrates remained deep in the oceans, causing yet another eruption of greenhouse gases into the atmosphere, making everything worse. This cycle would have happened not once but many times over a staggeringly long period—what we now call the Permian Extinction Event, which may have lasted as many as fifteen million years.

The wonder is that anything at all survived.

The Permian was the biggest, but not the first, mass extinction in Earth's checkered history. Two such episodes preceded it, and two followed it. Or possibly three, if you count the one we're in now.

The earliest of the major die-offs is called the End Ordovician, or Ordovician-Silurian Extinction. It seems really to have been a series, rather than a single event, and is usually attributed to sudden and repeated instances of extreme glaciation. It began about 488 million years ago, and then again about 439 million years ago. With a nearly 50-million year intermission between acts, you'd think these events would be counted as two plays, rather than as two acts of a single play—that's an awfully long time for the curtain to be down and the audience out buying popcorn. But partly because it was so long ago, and partly because the fossil record and therefore the evidence remain patchy, they are generally lumped together into one and are counted together as the second-largest extinction event in the planet's history.

Life did exist 439 million years ago, but not much of what you would recognize from visiting a zoo today. What few multicellular animals existed, like brachiopods, bryozoans, conodonts, and trilobites, generally lived in shallow seas. Early clams showed up at about this time, and the first primitive fish. Still, perhaps 60 percent of all marine genera went missing.

What caused all this? One early theory, proposed by scientists at NASA (whose preoccupation, naturally enough, was looking

upward), suggested a gamma ray burst from a star exploding somewhere within six thousand light-years of Earth. Even ten seconds of
such a burst would have been enough to strip away one of the planetary defenses, the ozone layer, thereby exposing surface-dwelling
organisms to sharply increased ultraviolet radiation. This would have
killed many species and caused a sharp drop in temperature, enough
to set off the glaciation that is known to have followed. Since this
version of the idea came from NASA, no one ridiculed it. But nor
did anyone find any evidence that such an explosion had ever
happened, and the theory has now been abandoned, or at least
shelved.

That's where it remains today. No one knows what set off the glaciation. All science can say is that several episodes of deep freeze succeeded
the previously balmy Ordovician epoch, preceded each time by a sharp
drop in atmospheric CO_2 levels. As Gondwana, the southern mega-

Dates of Mass Extinction Events

End Ordovician (Ordovician-Silurian)
488–439 Ma

Devonian Extinction Event
360 Ma

Permian Extinction Event (The Great Dying)
251 Ma

Triassic-Jurassic Extinction Event
200 Ma

KT (Cretaceous-Tertiary) (the end of the dinosaurs)
65.6 Ma

(Ma = million years from the present)

continent, drifted over the south pole, it began to freeze, and as the global temperatures dropped, the shallow seas typical of the era began to shrink. Repeated shrinking and rising over the eons eliminated many species each time, lowering the overall biodiversity.

Once again, though, enough life forms survived, and evolution resumed its rudely interrupted course.

The next interruption came at the end of the Devonian period 80 million years later, around 360 million years before the present.

This one, logically enough called the End Devonian Extinction Event, killed around 70 percent of all living creatures, and again no one really knows why. For some reason, animals living in the shallow seas were the least likely to survive, and that's all that is known, except that the Devonian extinction was definitely not a sudden event—estimates of how long the episodic extinctions lasted range from half a million to fifteen million years. Three million is the best guess—and some suspect that this was not a true extinction event at all, but rather one during which evolution took a holiday—speciation, as the term goes, seems to have diminished sharply at this period. Not so much a dying off, then, as a not creating anything new.

A more popular extinction (at least as these things are marked by the number of scientists studying them) was the Triassic-Jurassic Extinction Event, somewhere around two hundred million years ago, which paved the way for the ever-charismatic dinosaurs by eliminating their earlier rivals, the archosaurs, as well as the last of the great amphibians—after this, whales and elephants went their separate ways. More is known about this event than the one that preceded it. It happened just as Pangea was breaking up, was over in fewer than ten thousand years, and wiped out at least half the species known to have been living on Earth at the time.

It was definitely a homemade event. The extinctions were triggered by another series of massive eruptions, this time pushing huge quantities of lava out of the Central Atlantic Magmatic Province, the region around the modern Atlantic. This event had radical consequences, among them the opening up of the Atlantic Ocean, which had been a pond until then. Rocks from these eruptions are now found in many quite literally far-flung places—the eastern United States, eastern Brazil, North Africa, Spain.

As with the End Ordovician Extinctions, the eruptions caused the familiar cycle of early deep freezing followed by radical global warming, itself caused by volcanic CO_2 outgassing and, again, by the sudden melting of methane hydrates.

A nd then there was the death of the dinosaurs themselves. Hell Creek in New Mexico is arid scrubland, with great ridges of rocky outcrops, thorn bushes, stubborn grasses, and the occasional scrawny loblolly pine. It looks just like the countryside where I grew up, in South Africa near the Lesotho border, the same stark and beautiful mountains, the same smells, the same sounds, the same aridity but prone to sudden thunderstorms and flash floods. Pines instead of the acacia trees I remembered, rattlers instead of puff adders, but otherwise wonderfully familiar. Parts of Australia look like this, and South America and Afghanistan, but in one small way Hell Creek is special, though not quite unique, for near the top of the escarpment, so small you can hardly see it, just a half inch or so thick, is one of the most interesting and famous rock layers in the world, the so-called KT boundary.

The KT boundary marks the end of the Cretaceous Period and the beginning of what used to be named the Tertiary Period. (K is the usual abbreviation for Cretaceous, to avoid confusion with Carboniferous, which is abbreviated to C; Tertiary, for its part, is no longer used; the start of what used to be called Tertiary is now the start of the Paleocene epoch. Still, KT it remains.) The boundary layer can also be seen in the Badlands near Drumheller, Alberta, and at one or two places in Colorado, notably the Raton Pass and Trinidad Lake State Park.

It's famous both for what it signifies and for what it contains. If looked at really carefully, it seems to be a double layer. The lower, gray layer is thicker, maybe about 4 inches on average, though much thicker in places. The upper, darker layer, only about 0.2 inch thick, contains high levels of iridium, impact spherules, and traces of shocked quartz; this "iridium layer" is found worldwide. Above that is a thick layer of mudstone, shale, and coal, clear evidence that after the Event, whatever it was, this area was swamp.

By general agreement, what this boundary layer contains compellingly suggests a mechanism for the dinosaur die-off, the fifth of the Big Five Extinctions that have occurred on our planet since life began.

Agreement, but not consensus, as will become clear.

The story begins with a series of imaginative leaps by a scientist called Walter Alvarez, the son of Nobel winner Luis Alvarez, who had been working as an oil company geologist and a researcher in Italy. For a series of unconnected reasons, Walter had been studying the curiously rapid extinction about sixty-five million years ago of small and uncharismatic little creatures called forams—this was in the same period that the dinosaurs disappeared. He had noticed that these forams very nearly went extinct right at the KT boundary layer—on the upper layers of the Cretaceous they were as abundant and as big as sand grains; at the lower levels of the Tertiary only the very smallest ones survived. Why was this so abrupt?

At the boundary, the foram tipping point, was a layer of clay about a half inch thick, without any fossils at all—did the clay cause the extinctions? This enigmatic little layer, it turned out, contained anomalously large concentrations of a rare element called iridium—rare on Earth, since it is very heavy and would have sunk into the core early in Earth's history, but far from rare in meteorites.

That was Walter's eureka moment. This iridium layer could, he suggested, have been laid down by a massive asteroid impact. . . .

This was a huge intellectual risk. Conventional paleontological wisdom at the time was still that the dinosaurs had expired of their own evolutionary incompetence—they were too big and stupid to survive the coming of, well, us. This charmingly anthropocentric view is typified by the 1886 comment by Alexander Winchell, professor of geology at the University of Michigan, who put the end of the dinosaurs this way: "A higher type is now standing at the threshold of being. A knell is sounding the funeral of the reptilian dynasty. The saurian hordes shrink away before the approach of a superior being. After a splendid reign, the dynasty of reptiles crumbles to the ground. . . ."[3] Even after the mid-twentieth century, many geologists were still imprisoned in the Uniformitarian view bequeathed to their science by Charles Lyell, and while they accepted homegrown catastrophes caused by earthquakes and volcanoes, they were still reeling from the popular acclaim given to the worlds-in-collision theories of Immanuel Velikovsky, of whom more later, and couldn't accept errant rocks from space. And so this notion of an asteroid-caused catastrophic extinction was, still, the purest heresy.

In June 1980 Walter published his team's results in the journal *Science* and delivered an oral presentation to a meeting in Denmark. But at the last minute he thought better of his notion to float the idea of an asteroid impact and reported only on the iridium anomaly. It turned out that a Dutch scientist, Jan Smit, had found similar anomalies elsewhere, and in the next few years scientists fanned out all over the world to test for iridium wherever they could uncover the KT boundary, most of them hoping to shake the Alvarez theory, which had by now become the Alvarez Impact Theory. Instead, they kept finding KT samples with the same elevated iridium levels the Alvarez team had found, levels that corresponded very neatly with the time the dinosaurs vanished, and the notion that major natural disasters may have had extraterrestrial origins began to enter the scientific mainstream.

But where was the smoking gun? If there had been an asteroid impact, where was the crater? A Canadian geologist, Richard Grieve, compiled a list of possible impact craters, but there were very few, or at least very few still visible. Meteor Crater in Arizona was too small and too young. The Manson Crater in Iowa was suspected for a time, but it too was too small (its impact crater, now buried deep under glacial till, is about 24 miles in diameter, and is also too young, about 74 million years old). Sudbury, Ontario, and Vredefort, South Africa, were possibly the right size, but much too early. From the amount of iridium laid down, the impactor must have been a good 6 miles in diameter, about the size of Manhattan, and would have struck with the force of 100 trillion tons of TNT, or about 2 million of the largest nuclear devices ever exploded. The crater must be at least 62 miles in diameter, and probably much more.

Finally a prime candidate was found, in the Gulf of Mexico on the Yucatan, half on land (though long smoothed over) and half under the gulf's water. A Canadian grad student, Alan Hildebrand, identified it as a suspect and published his work in the journal *Geology*,[4] working from evidence presented by Tony Camargo and Glen Pentfield, geophysicists working for the Mexican government.

The theory, as it emerged, was this: the asteroid, called Chicxulub, Mayan for "tail of the devil," a child of the Baptistima asteroid family,[5] struck the Yucatan at a shallow angle, maybe 20 to 30 degrees, traveling northwest, spraying its debris well into Alberta. It vaporized a massive chunk of the Earth's crust on impact, blasting up a plume of material more than a hundred kilometers wide that rose above the atmosphere into space, before plunging back to Earth. A shock wave plowed through the crust into the granite beneath, and a second shock wave tore the projectile itself apart. The limestone beneath the impact ignited, causing a second huge fireball, ejecting massive quantities of CO_2. This was followed by a gigantic tsunami, maybe a half mile high. Everything within 1,250 miles of the impact would have been destroyed. Gigantic fires would have been ignited in the superheated atmosphere, raised to temperatures that would have been four times that of the sun. The world's forests went up in smoke, the shallower seas boiled, and billions of tons of debris were ejected into the air. Photosynthesis stopped and the food chain collapsed, killing off the alpha species, the dinosaurs. When the fires went out, the Earth remained dark and plunged into a deep freeze, followed not long afterwards by runaway greenhouse effects. It would have taken a hundred thousand years for equilibrium to be restored. By that time 16 percent of marine families had died, and 18 percent of land families—maybe as many as 50 percent of all species—perished.

It was worst in the northern hemisphere, particularly in the Americas. As many as 75 percent of plant species died, followed by a spike in the growth of opportunistic ferns; in the southern hemisphere, there were mass kills too, but population collapses were shorter-lived. Animals that depended on photosynthesis, like vegetarians, did badly.

Omnivores and insectivores survived rather well; so did the insects themselves. Placentals, our ancestors, did fine. All non-avian dinosaurs were killed. So were some other species, such as American marsupials. In the oceans and rivers, species that relied on living matter suffered, those that fed on detritus did quite well.

There is quite a lot of agreement about this scenario, but not complete agreement.

For example, no one doubts that Chicxulub exists, but then so do other candidates. More recently more craters dating from the same period have been found—the Boltysh Crater in Ukraine, 15 miles in diameter; the Silverpit Crater in the North Sea, 12 miles; Eagle Butte in Alberta, 6 miles; and Vista Alegre in Paraná State, Brazil, 6 miles. None of these is large enough to have done the damage, but it suggests either that Earth could have been bombarded by multiple impacts or possibly by a single large asteroid broken into several chunks.

Walter Alvarez, who seems to have an impish sense of humor (he titled his memoir of his discoveries *T. Rex and the Crater of Doom*), told a meeting of the American Geophysical Union in 1990 that "a few years ago, our problem was that we didn't have any craters to point our fingers at, whereas now we've got several. So now, maybe instead of a smoking gun we've got a smoking firing squad."[6]

But some scientists still believe that India's Deccan Traps, a phenomenon similar to, though smaller than, the Siberian Traps, actually caused the extinction—the Deccan eruptions poured out some 0.6 million cubic miles of basalt rock, happened at about the right time, about sixty-five million years ago, and are thought to have lasted about a million years. Just as the Siberian eruptions caused dust and sulfur gases that blocked sunlight, followed by increased CO_2 and methane that caused dramatic warming, so too would these eruptions. Even Alvarez agrees that, at the least, there were features in the pre-impact world that would have exacerbated (and prolonged) the asteroid's effect. For example, a very large crater, the Shiva Crater, has been discovered on the sea floor off India that dates to the right time and could have set off the Deccan eruptions. Shiva is believed not to have been an impact crater but a massive sinkhole, but it would still have had a precipitating effect.

At the same time as all this was going on, sea levels dropped precipitately from their Cretaceous levels, in an event called the Maastrichtian Sea-Level Regression. No one is sure why this happened, but one possible explanation is that for some reason the mid-sea ridges stopped their activity—stopped oozing new crust into the open—and instead sank from their own weight, dropping sea levels dramatically. This would have reduced the Earth's albedo, with consequent global warming and climate change.

Other paleontologists accept the asteroid impact but dispute the extinctions. Possibly, as Alvarez partisans have pointed out, this is because many paleontologists still lean toward gradualist theories of change, whereas astronomers and physicists lean toward catastrophic explanations.

The most prominent critic of the Chicxulub impact theory is Gerta Keller, professor of geosciences (paleontology) at Princeton University. For years she has maintained that the impact actually happened three hundred thousand years before the boundary itself, based on evidence she had collected in Mexico and Texas, and that the extinctions were much more gradual than supposed. In contention was a 33 foot layer laid down on top of the crater debris, which Keller maintains was laid down much too slowly to have been a result of catastrophic change. At best, she maintains, the Chicxulub impact would have stressed the ecosystems on Earth, reducing the number of species. It is also possible that the 33 foot layer was tsunami-produced and therefore contemporaneous with the impact.

Another critic worth listening to is Robert Bakker, the man who had proposed, to much derision, that the dinosaurs weren't cold-blooded lumbering creatures and were not extinct at all, but had evolved into birds. This notion is now widely accepted. But, in *The Dinosaur Heresies*, Bakker argues against the asteroid impact theory.

His solution is beguilingly simple. As the sea levels dropped in a cooling world, two things happened: most of the marine life that existed in warm waters would have been driven into deeper and colder waters, where many perished; at the same time, new land bridges appeared, which in turn meant the sudden appearance everywhere of new invasive species, with new varieties of predators against which the natives were unprotected. These predators, he pointed out, would not have been just large toothed creatures, but would generally have been microbially tiny: bacteria.

Then, early in 2007, a Smithsonian expedition recovered a deep-sea drill core that dramatically confirmed the impact theory, clearly showing dust and ash fallout from the explosion and the consequent devastation of marine life. This core was recovered 298 miles east of Florida, almost 1,250 miles from the Yucatan.

The debate is ongoing.

One way or the other, the dinosaurs died, except for the birds. I always liked the cartoon that depicted two battery chickens in a cage, watching a guy with a cleaver approaching. One mutters to the other, "Just think, once we were 30 feet tall, with razor teeth and awesome claws. Oh for the old days."

A cliché of the dinosaur extinction debate, regardless of how the extinctions happened, was that the disappearance enabled the worldwide proliferation of mammalian species, which until then had been small creeping things of no particular importance and of low fecundity. This has now been proved wrong.

A paper published in the journal *Nature* in 2007 argued, to the contrary, that the evidence showed that most of the mammalian proliferation happened about ninety-three million years ago, while the dinosaurs were still dominant, in what the paper called "a temporal hot spot of diversification." Then the rate of speciation fell, remained low during the extinction events, and spiked again only well after the dinos had disappeared. "Therefore the demise of the non-avian dinosaurs, and the KT mass extinction event in general, do not seem to have had a substantial direct impact on the evolutionary dynamics of extant mammalian lineages."[7] Most of the mammalian species produced in the first spike themselves went extinct in the KT event. It wasn't until another such hot-spot spike, about fifty-five million years ago, that most of the recognizably modern mammals emerged. The reason: a more moderate, more "modern" climate. It turns out that the average "species lifespan" for mammals is about 2.5 million years. After that amount of time, it seems, a species' time is pretty well up, catastrophes or not.[8]

And lest you are comforted by all those zeroes after the numbers—after all, most of these events happened many millions of years ago—and like to believe that we are now in some version of an epochal safe house, increasing evidence suggests that at least one cosmic collision happened well within human history.

For years scientists have puzzled over the Clovis people, named after Clovis, New Mexico, where artifacts from their culture were first found. Also named paleo-Indians, they were generally agreed to be the first human inhabitants of the Americas—or at least no evidence for pre-Clovis cultures has been found. They either crossed the Beringia land bridge over the Bering Strait from Siberia to Alaska during the period of lowered sea levels during the ice age or they made their way in some early version of the shallop along the shores of the Pacific. The culture lasted for about half a millennium and then vanished.

The conventional wisdom is that they were responsible for the extinction of several large fauna types in the Americas, most notably the mammoths, which disappeared at about the same time. Other species that went missing in the period include short-headed bears, ground sloths, and North American camels. The only real artifacts recovered from Clovis settlements, in fact, were many hundreds of distinctive fluted stone spear points, called, naturally enough, Clovis points. These were supposed to be evidence that the Clovis people were skilled hunters. Recent evidence has cast some doubt on this cozy hypothesis (the "Pleistocene Overkill Hypothesis"), which was much beloved by those who believe in humankind's relentlessly negative impact on the planet. A report in the *Journal of World Prehistory*, however, suggests something different. Its authors, Donald Grayson of the University of Washington and David Meltzer of Southern Methodist University, put it this way:

Of the 76 localities with asserted associations between people and now-extinct Pleistocene mammals, we found only 14 (12 for mammoth, two for mastodon) with secure evidence linking the two in a way suggestive of predation. This result provides little support for the assertion that big-game hunting was a significant element in Clovis-age subsistence strategies. This is not to say that such hunting never occurred: we have clear evidence that proboscideans (mammoths and mastodons) were taken by Clovis groups. It just did not occur very often.

So what, then, caused the mammoths and mastodons to disappear—along with the Clovis people?

Some archeologists suggested that climate change had an impact, since both disappearances happened at the start of what is now called the Younger Dryas Cold Climate Period, a 1,300-year stretch more coarsely called the Big Freeze, which occurred, perversely, at the time the globe was still warming from the previous ice age. Now the impact is increasingly being seen as caused by, well, an impact. The theory was first floated in May 2007 by a team of archeologists from several American universities, reporting data from many sites, including some off the California coast and some in the Great Lakes region. The scientists include another father and son team, Douglas Kennett of the University of Oregon and his father, James, of the University of California at Santa Barbara.[9]

It is generally known that the Younger Dryas was set off by a reversal in the world's ocean currents; the new theory accepts this but suggests that the reversal was itself set off by an extraterrestrial impact, probably a comet, that exploded somewhere above where the Great Lakes are now, melting massive quantities of ice and causing millions of tons of fresh water to pour into the Atlantic, pushing the currents into reverse and halting the thermohaline circulation.

A notable countervailing theory comes from paleoclimatologist Jeff Severinghaus, whose study of ice cores persuades him that the cooling began before the putative impact, but nevertheless the Kennetts reported that, in many places in the United States and Canada, evidence for an impact is easy to find, merely by digging down far enough (and not that far—the thing happened "like yesterday," James Kennett says), where a few inches of charcoal will appear, evidence of wildfires that spanned the continent after an object about a 0.6 miles across broke up and dissipated in a fiery plunge. "The highest concentrations of extra-terrestrial impact materials occur in the Great Lakes area and spread out from there," Douglas Kennett said. "With major effects on humans . . . shockwaves, heat, flooding, wildfires." The charcoal layer includes other evidence of an impact—metallic microspherules, glass-like beads made of carbon, tiny diamonds the size of a micron, and other substances. Not much iridium, though, which suggests a comet rather than an asteroid, since comets are mostly muddy ice. No crater has been found, but if the impact was on the massive ice sheets, none would be expected. Other impact markers have been found at twenty-six sites, ranging from California to Belgium.

Not all large mammals were wiped out, and some evidence that supports the theory can be teased out from those that did survive. Bison, for example, survived into modern times, but they are not like the bison the Clovis people would have hunted. They are smaller, for one thing, and they are also remarkably alike in their genetic makeup, suggesting that they descend from a very small group of animals.

Were the first Americans also wiped out? As a culture, the Clovis people disappeared and were succeeded by the Folsom culture, also defined by its arrowheads. It is not yet known whether the Folsom people grew from Clovis survivors, or whether they were an entirely new immigration from Asia. If the latter, then America was, for a few centuries at least, as bereft of people as of mammoths.

Two more things to ponder: a 1998 survey by the American Museum of Natural History found that 70 percent of biologists view our own time as part of a "mass extinction event," similar to the Permian and the KT, and possibly one of the fastest ever. For exam-ple, Edward O. Wilson, who was in early 2007 the Pellegrino Research Professor in Entomology at Harvard, predicted that human-driven destruction of the biosphere could cause the extinction of one-half of all species in the next hundred years. Wilson is also a Fellow of the Committee for Skeptical Inquiry and a laureate of the International Academy of Humanism, and some of his contemporaries are skeptical of his objectivity. But it's a sobering thought, nonetheless: the human species as a blundering asteroid.

And lastly: life nearly went extinct several times. So did humankind. But both have triumphantly survived.

PART THREE

Peril by Peril

SEVEN

The Perils Without:
Comets and Asteroids

It Came from Outer Space
—1953 sci-fi movie in 3-D, starring Barbara Rush and directed by
Jack Arnold

Out of somewhere, out of the Deep Dark of the great void past
Pluto, came . . . something. A planetesimal, perhaps, a failed
planet fragment, or the ejecta of some cataclysmic explosion somewhere far away in the cosmos, no one knows, a rocky, metallic *something*. It hurtled through the void for a long, long time, and then it
came . . . here.

Number 5 shaft of the Creighton Mine in Sudbury, Ontario, goes
straight down about 3,800 feet, deep into the sweltering Earth.
The hoist cage, which looks like but really isn't an antique, is now used
mostly for transporting goods, but thirty years ago it was the latest thing
in vertical people movers, an open cage that slid at frightening speed
down the sweating rock. I wore heavy gumboots, a slicker, and a hard
hat that fit ill over my ears and tipped my glasses skyward, and was
treated with a mixture of solicitousness and contempt by the miners

who did this ride every day. (Most of their contempt, I remember, was reserved for the management type who insisted on accompanying me that day; as a mere newspaper reporter, I really wasn't worth condescending to, there not being a strike on that month.) That 0.6 mile down was the bottom of the world, then. Now, at 3,900 or so feet you have to move over laterally a few hundred feet to wait for the main cage to descend Number 9 shaft, to take you down farther, much farther. Push the down button and Number 9's cage will descend another 3,300 feet or so, coming to rest, finally, 6,800 feet below the surface, more than 1.3 miles down, the deepest mine in the western hemisphere. (The East Rand gold mine in South Africa is the world's deepest, at 11,762 feet, at which level the rocks are upwards of 122 degrees Fahrenheit, and the icemaker to keep the whole thing cool produces about 20,000 tons of ice every day.)

In the hundred or so years this mine and others in the Sudbury basin have been operating, more than 19 million tons of nickel and copper have been hauled to the surface, processed, and sold, essentially spreading Sudbury thinly over every country in the industrialized world and beyond. That pile of heavy metal would be worth somewhere around $350 billion at today's prices.

A gift, if that's what it was, from deep space.

Nothing or nobody would have seen it coming. There was life on Earth then, somewhere around two billion years ago, but nothing conscious. And even if some dim creature had been aware, it could have done nothing to escape. The object, whatever it was, would have been dark until it was lit up by the heat of atmospheric entry, but by then it would have been far too late even to duck. A second later it hit the ground at somewhere around 89,500 miles an hour, ripping a 124-mile crater 12 miles into the ground, melting more than 6.5 cubic miles of rock, hurling a cloud of debris through the stratosphere that spread around the globe, blocking out the sunlight for decades, and laying a thick cloud of ash over the continent, the not-yet-North America.

It was so big, maybe 6 miles across, the same size as the impactor that later destroyed the dinosaurs, that the nanosecond it first touched the ground the other side would still have been as high as the peak of Mount Everest. Its volume would have been equivalent to all the buildings in North America. The energy the impact released would have been similar to Chicxulub's, about 10,000 times the world's entire nuclear arsenal, or about 100 million megatons of TNT.

In the decades and centuries after the impact, as the melted rock cooled, heavy nickel and copper settled to the bottom of the crater. In the eons to follow, the surface was gradually worn down by erosion and by the insistent pressure of tectonic drift, which caused great cracks in the surrounding rock and squeezed the whole basin into a crude oval. Some deeper rocks—and the metals they contained—were squeezed closer to the surface.

I'd been sent to Sudbury by a magazine editor who wanted me to write a story for which she had already written the headline: Who Killed Happy Valley? How could she resist a name like that, or photographs of the desolate wilderness that Happy Valley (a suburb of Sudbury) had become? I still remember the Inco mine manager virtually rolling his eyes when I walked into his office with my preconceptions firmly in place—it was why he sent me down below, hoping the miners would insert some countervailing fact into my journalistic skull.

Well, Happy Valley *was* a wasteland. Sudbury's meteor crater is rich in many minerals—nickel, copper, platinum, and many others, but is also rich in sulfur, and when the nickel-copper ore was smelted, the sulfur simply drifted into the environment, where it settled on the vegetation like a toxin. Much of it drifted higher into the atmosphere, where it combined with aerosols to form acid rain, which in its turn contaminated much of the region. The problem was "solved" by building the Inco Superstack, thrusting the sulfur higher into the air, thus removing the pollution problem from Ontario to, say, Nova Scotia and beyond, even as far as Sweden. It didn't help that much of the area had been stripped of its trees in years past, by loggers; Sudbury wood helped rebuild Chicago after the Great Fire of 1871, and the place hadn't really regrown a forest cover; what soil remained was black and charred, as though it were the entrance to Hades.

Happy Valley, happily for me, was downwind from the smelters and therefore thick with pollutants, and I was able to satisfy the editor by finding an old Italian who lived in a shack in the desolate landscape but who had carefully cultivated a little patch of tomatoes, the only color for almost a mile, making a suitably poignant if rather biased picture.

In the 1960s the Sudbury area was visited by astronauts training for the Apollo manned lunar mission, mostly to become familiar with so-called shatter cones, a characteristic rock formation found near meteor impact craters. Inevitably, the popular misconception, notably fueled by articles such as mine, was that the astronauts came to Sudbury because it most closely resembled the lifeless moonscape they were to be visiting. So it's worth noting that in 1992 the city of Sudbury was given a Local Government Honors Award by the United Nations, in recognition of its astonishing re-greening efforts. Now the city is known mostly for its lakes, there being some three hundred within its municipal boundaries, including the largest lake anywhere entirely contained within a city, Lake Wanapitei, which is 4.7 miles in diameter.

Lake Wanapitei is also a meteor impact crater. But it is unrelated to the Sudbury strike that caused Big Nickel, being only thirty-seven or so million years old. Sudbury hasn't been a lucky place, that way.

Sudbury's certainly wasn't the only, or even the largest, planetesimal to slam into the Earth. At least 150 impact craters are known, most of them for obvious reasons of identification, on land. Because the Earth is mostly water, many more would have landed at sea.

The largest, although not the most famous, is the Vredefort Dome, an enormous planetary wound, or astrobleme, to use its unexpectedly euphonious technical description, some 236 miles in diameter. Vredefort is a small town southwest of Johannesburg, in South Africa. I once saw its dome when I was a kid—my father had hauled the family to the nearby town of Parys in a futile search for a long-lost relative—but had paid it no mind. It didn't look at all like a dome after two billion years, rather resembling the other kopjies, or small hills, that dotted the Transvaal and my native province of Orange Free State. It was surrounded by a depression that could have been a crater, but it was not easy to see. Still, the Vredefort Dome is the remnant of the world's greatest known "energy event," which is thought to have caused devastating global changes, very likely changed the course of evolutionary history, and has fascinated geologists ever since it was discovered, yielding as it does the only known example of a full geological profile of an astrobleme below the crater rim.

More famous than Vredefort, and certainly more famous than Sudbury, is the crater that straddles the coastline on the Mexican Gulf near Chicxulub, which, as reported, is intimately associated with the extinction of our predecessor alpha species, the dinosaurs.

Others include Manicouagan, in Quebec, at 62 miles; Popigai, in Russia, also 62 miles; Acraman, in Australia, at 56 miles; Chesapeake, in the Washington, D.C. area, at 53 miles; Montagnais, almost within view of my house, just off the continental shelf in Nova Scotia, at 28 miles; and St. Martin, Manitoba, at 25 miles. Montagnais is about 50 million years old—give or take a generation or two, well before my time.

The Barringer Crater (formerly known as the Canyon Diablo Crater), near Winslow and Flagstaff, Arizona, is much younger, about 50,000 years old. The impactor was probably only about 131 feet across, but it punched a 656-feet-deep hole in the desert, maybe 394 feet in diameter, still surrounded by a jumble of smashed boulders the size of houses. The explosion that followed its impact was a trivial thing by planetary standards. With the force of no more than, say, 150 atomic bombs or so, it would have caused only local damage.

Among the most picturesque crater impact zones are the lakes in Quebec named Lacs à l'Eau Claire, or Clearwater Lakes. Actually, there are 25 lakes in Quebec with that name, but these twins, 16 and 22 kilometers in diameter, the result of a binary asteroid impact, are the most northerly, on the latitude of Hudson Bay. The impact that caused them occurred probably around 290 million years ago.

In case you think that these showers of planetary fragments are now over, and that our cosmic neighborhood has settled into a genteel senescence, consider this:

About nine hundred years ago, a laughable interval in cosmic time, on June 25, 1178, five English monks actually watched an asteroid striking the moon. Of course, they didn't describe it thus, but as a "flaming torch" that caused a dramatic display of flame and spark. The whole event was chronicled for posterity by Gervase of Canterbury, the same Gervase who had recounted the death of Thomas à Becket, murdered by Henry II and immortalized in due course by Chaucer in the Canterbury Tales.

> This year on the Sunday before the Feast of Saint John the Baptist, [Gervase recounted], after sunset when the moon was first seen, a marvelous sign was seen by five or more men sitting facing it. Now, there was a clear new moon, as was usual at that phase, its horns extended to the east; and behold suddenly the upper horn was divided in two. Out of the middle of its division a burning torch sprang, throwing out a long way, flames, coals and sparks. As well, the moon's body which was lower, twisted as though anxious, and in the words of those who told me and had seen it with their own eyes, the moon palpitated like a pummeled snake. After this it returned to its proper state. This vicissitude repeated itself a dozen times or more, namely that the fire took on tormented forms variously at random, and afterwards returned to its prior state. Even after these vicissitudes, from horn to horn, that means along its length, it became semi-black.

The monks' tale was widely disbelieved, but Gervase declared that "they were prepared to stake their honor on an oath that they had made no addition of falsification" in their narrative.

In 1976, an astrophysicist, Jack Hartung, asserted that the crater we now call Giordano Bruno was probably caused by the monks' meteor (others have disagreed, suggesting that the monks might have seen a transiting comet). Hartung has also pointed out that what we now call the Little Ice Age began only a few years after the Canterbury impact, and that the two phenomena could be related. To this degree we are influenced by our little satellite.

One more interesting observation: In the late 1970s, instrument packages on the surface of the moon showed that the entire satellite was still vibrating like a huge bell, with a period of about three years, which is what you would expect a mere eight hundred years after a major asteroid impact.

Such impacts on the moon are not rare, either. By late 2007, NASA's Meteoroid Environment Office at the Marshall Space Flight Center had recorded sixty-two impacts in two years, some of them causing enough light to be seen through ordinary backyard telescopes.

In 1807, on December 14, fragments of a large meteorite crashed near Weston, Connecticut, raining rocks in and around the township. The inhabitants were "strongly impressed with the idea that these stones contained gold and silver, [so] they subjected them to all the tortures of ancient alchemy, and the goldsmith's crucible, the forge, the blacksmith's anvil, were employed in vain to elicit riches which existed only in the imagination." This, sardonic language and all, was part of a report by Benjamin Silliman and a colleague, who were sent to the area from Yale University to investigate. The report was read to the American Philosophical Society, where it was well received, except by Thomas Jefferson, who said afterwards, "I would more easily believe that two Yankee professors would lie than that stones would fall from heaven."[1]

And then on June 30, 1908, a fireball exploded near the remote village of Tungusta, in Siberia. It was so remote that although rumors of the event circulated around Russia for years, it wasn't until 1927 that an expedition under Leonard Kulik actually reached the site. They found an area of devastation about 800 square miles in extent; at the center, the trees were upright though stripped of bark, and beyond those all were flattened outward. Until recently, no one had found any impact crater, and so cosmologists speculated that the impactor would have been around 165 feet across and would have vaporized about 4 to 5 miles above the surface, exploding with the power of a 50-megaton weapon, causing an atmospheric shock wave that circled the globe. In 2007, though, a team from Italy suggested that a funnel-shaped nearly circular lake, called Lake Cheko, may be a crater left by a remnant of the explosion. The lake is about 5 miles away from the epicenter.

Closer to our own time, a mere fifty years ago, a woman in Sylacauga, Alabama, sitting peaceably in her own living room, was slightly injured by an 11-pound meteorite that crashed through her roof. In April 1971 another meteorite caused minor damage to a house in Wethersfield, Connecticut, and eleven years later another damaged another house in the same town, only a half mile away.[2]

Even more recently, during the week of July 16 through 22 of 1994, a series of comet fragments crashed into Jupiter. Several of these twenty chunks of ice and stone were 3 miles in diameter, as big as a small mountain, and they created wounds in the Jovian surface the size of the Earth, ejecting billions of tons of material high above the atmosphere.

They're still out there. And they occasionally come to get us. In March 2007 a planetary defense conference was summoned by the U.S. Congress, whose attention had been caught by the spectacle, seen live on TV, of a comet wreaking havoc on a planet much greater by far than Earth, namely Jupiter. This was a serious conference attended by serious people, among them NASA, the American Institute of Aeronautics and Astronautics, the Japan Aerospace Exploration Agency (JAXA), the Indian Space Research Organization (ISRO), the European Space Agency, and a number of physics and astrophysics labs.[3] An excerpt from the White Paper produced afterwards said this: "It is well known that there have been impacts of large Near Earth Objects (NEOs) in the past history of Earth. It is also well known that, while the probability of impact in the next month or year is small, impacts of objects large enough to seriously modify or potentially end life, as we know it, are inevitable."

It would cost about a billion dollars, the participants calculated, to find and catalog, by around 2020, some 90 percent of the more than 20,000 asteroids large enough to be potential planet killers that regularly cross Earth's orbit.

This didn't seem like such a large amount, with the war in Iraq costing more than that a week, and considering the odds the assembled scientists then calculated: the probability of a "dinosaur-killer" impact—that is, one that would very possibly end life as we know it—is about 1 in 1 million this century. The probability of a civilization-ending impact is rather larger—a bit less than 1 in 1,000 this century. For a smaller, Tungusta-class impact, near the lower size for penetration of the atmosphere, but still large enough to destroy a city, the odds are higher still: maybe one chance in ten that we'll get a hit this century.[4] If you look at it in terms of expectations rather than probabilities, we could expect a planet buster every 250,000 years, and a regional disaster like Tungusta's about every 100 years. Those are not comforting odds.

Of course, "impact frequency" is misleading—see the discussion of probabilities in Chapter 2; neither asteroids nor comets hew to any kind of schedule. A testy Phil Plait, who runs a hectoring website called Bad Astronomy, puts it this way: "Asteroid and other impacts are a stochastic process. They're random. We could get hit tonight by a big one, or it might be a million years from now. Impact frequency is a statistical one only, so we can't be overdue for a hit. That's like a gambler saying, I've lost six times in a row, so I'm overdue for a win."

It is in any case impossible to catalog all the tiny Tungusta-sized asteroids. There are millions of them, for one thing, maybe a hundred million, all of them moving in unique orbits, all of them potential collisions with Earth. And you can spot them only by sheer luck, if they happen to transit another larger body that just happens to be under scrutiny by an astronomer at the time. In the case of several recent near misses, they were seen only after they had whizzed by. In 1989 a three-quarter-mile-long asteroid was discovered only after it had crossed Earth's orbit at a spot we had been only six hours earlier. In May 1996, another, roughly the same size, was discovered only four days before it sped across Earth's orbit, missing the planet by four hours.[5]

Mostly, we'd get no more warning than the Sudbury basin got when its load of nickel and iron came crashing down. Maybe a second or two, no more. Then it would be all over.

From the point of view of celestial mechanics, comets, especially long-term comets, are the rural cousins of asteroids. They can come into town (if in our hubris we define our own neighborhood as the solar system's downtown) from any direction at any time, but almost always from the distant Oort cloud, which as we have seen stretches partway to the next star system. Asteroids, on the other hand, usually jostle each other much closer in and hang out in gangs at the mall somewhere between Jupiter and Mars.

At least one bright comet pops up in our part of the solar system every five to ten years. Shoemaker-Levy, which immolated itself by banging into Jupiter, was one such. Two others were Hyakutake and Hale-Bopp. Hyakutake was merely average in size, but came as close as 0.10 AU (9.3 million miles) from Earth, which made it appear especially spectacular. Hale-Bopp also looked bright, though it was a good deal farther away, 1.32 AU (122 million miles) at its closest; it looked bright because it was very big, ten times the size of the more famous Halley.

Neither of those had any real prospect of colliding with us. But that's not quite as true for Comet Swift-Tuttle, which passed through our neighborhood in 1992. The best estimate for the time of closest passage to the sun during its next approach is in 2126. It does pass through Earth's orbit, and a slight deviation in its track (a mere 15 days in 120 years or so) would mean that the comet could collide with the Earth on August 14, 2126. Hmmm.

Before 1992, the comet's previous appearance was when Abraham Lincoln was president of the United States, in 1862. It was spotted and named by two American astronomers, Lewis Swift and Horace Tuttle, who calculated its orbit as 120 years. It should have come back, then, in 1982. But it didn't.

For an explanation of why it was late, we have to go back to an Italian astronomer named Giovanni Schiaparelli, whose telescope was pointed at the sky during Hale-Bopp's last bypass. Schiaparelli was a meticulous observer, though he later became embroiled in controversy when he claimed to have spotted ditches on Mars he called *canali*, an unfortunate name that soon gave rise to speculations that they were in fact irrigation aqueducts of some long-vanished Martian civilization. This was a speculation that Schiaparelli tried in vain to suppress, without success. He had better luck chasing the comet.

Shortly after its appearance (or reappearance, as it turned out) Schiaparelli noticed that its orbit was very close to that of the dust particles that make up the Perseid meteor shower each August. During the Perseid peak, hundreds and sometimes thousands of meteors an hour flash through the night sky. Schiaparelli's theory, now widely accepted, established comets as the originators of meteor showers—as comets move close to the sun, the intense heat gain explosively melts the ice, the resulting water jetting away from the surface, pulling some of the comet's dust out with it. When the Earth crosses the orbit, at the same time each year, it plows through this cometary debris, unleashing a meteor shower.

Through the 1970s, the number of meteors seen each year in the Perseid meteor shower increased, and it seemed that their parent, Swift-Tuttle, was about to reappear. But it failed to show, and soon afterwards, Perseid meteor activity dropped sharply. Perhaps the comet had somehow come and gone, unnoticed? Or had it disintegrated?

Then, in 1973, Brian Marsden, director of the Minor Planet Center at the Harvard-Smithsonian Center for Astrophysics, suggested that the comet seen in 1862 might be the same one reported in 1737 by a Jesuit missionary, Ignatius Kegler, in Beijing, China. Possibly those sun-caused cometary "jets" had acted like a rocket exhaust and had slightly altered its orbit? If this were true, Marsden calculated, Comet Swift-Tuttle would now have a period of 130 years, instead of 120, and would return at the end of 1992. Indeed, increased numbers of Perseid meteors in the early 1990s indicated the comet might be near, and on September 26, 1992, an amateur Japanese astronomer, Tsuruhiko Kiuchi, using six-inch binoculars, noticed a comet moving through the Big Dipper in an area where scientists had calculated Comet Swift-Tuttle should be. Soon other astronomers confirmed that the lost comet had been found.

Thus Marsden's 1973 prediction was confirmed, although the precise date was off by seventeen days. On November 7, 1992, the comet passed 110 million miles from Earth (its closest approach).

Armed with new observations, Marsden revised his calculations of its orbit and predicted the next perihelion would occur on August

14, 2126, 134 years distant. But if the actual date of the perihelion was off by 15 days from his prediction (as the 1992 perihelion had been off by 17 days), the comet and the Earth might be in the same place in space at the same time. Since Swift-Tuttle is about 6 miles across, about the same size as the asteroid that slammed into Sudbury, ruining most of North America and radically changing the global climate, and the same size also as the asteroid that probably wiped out the dinosaurs, it was worth paying attention.

In the years since, astronomers have tracked the comet back to much earlier observations, even to 188 AD, and now think the orbit is more stable than Marsden's calculations showed. Marsden himself has withdrawn his warnings.

And in any case, a collision would be fluky. The comet will be moving at a speed, relative to the Earth, of 37 miles per second, which means that it would have to occupy the exact same space for no more than a few minutes (out of 130 years) for us to get unlucky. An error of only one hour in its timing would mean it would miss by 62,000 miles.

And for this, we'd have plenty of warning.

Asteroids, though, are very different creatures. They are closer, for one thing, and would give us less time to react to a threat. There are many more of them, for another. Daily, about a hundred tons of interplanetary stuff drifts down to the Earth's surface. Most are tiny dust particles released by comets as their ices vaporize. The bigger pieces come from collision fragments of asteroids that have run into one another some eons ago.

Some of the bigger pieces floating around Out There are very large. Not moon-sized, but large enough, maybe a thousand kilometers— quite enough to melt down our planet.

As we have seen, they lurk in the sky between Mars and Jupiter, violent remnants of the solar system's early history. Nine out of ten meteorites are chondrites, containing tiny spherical nodules known as chondrules (as well as a lot of cosmic crud, including dust and water; the rest contain mostly nickel and iron). Most of these have broken off the remains of more than sixty different small planets. A few are even Martian rocks, blasted into space by massive impacts.

For planets to have accreted to their present size, they would have to have been bombarded with such smaller objects over millions to billions of years. And of course it still happens, as we have seen—with effects ranging from colorful fireballs to mass extinctions. Asteroids continually collide with each other as well. A recent report in the journal *Nature* tracked just such a collision, fortunately in the distant reaches of our solar system. It resulted in the breakup of a planetoid, some twenty-five kilometers in diameter, into a number of small but still substantial fragments, and has been conclusively dated to 5.8 million years ago, so astonishingly recent that humanoids walked our planet not long afterwards.[6]

A few worried flurries have emanated from scientists about asteroids in the last half dozen years, as more and more scientific attention is focused on their potential hazards.

The Near Earth Object Information Centre, based at the U.K. National Space Centre, issues fairly regular bulletins on errant objects with the potential for doing us damage. In 2003 the asteroid labeled 2003 QQ47 was a brief scare, before the all-clear was issued a few weeks later. Other minor flaps included 1997 XF11, which will make a near miss in 2028; 1999 AN10, an asteroid less than a mile wide that could pass particularly close to the Earth on August 7, 2027 (missing by just 23,000 miles); 2002 MN, a 325-foot rock that whizzed by only a third of the moon's distance away and that was only spotted three days after it had already passed; 1994 XM1, a little thing, only about 33 feet across, that came even closer on December 4, 1994; and 2002 NT7, a large object described as "the most threatening so far," which later missed by a long way (though it could still hit when it returns on February 1, 2060).

The two most interesting recent cases were the asteroids labeled 1950 DA and 2004 MN4, also called Apophis, after the Egyptian spirit of evil and destruction, a demon that was determined to plunge the world into eternal darkness, which seems appropriate enough.

NAME	APPROACH DATE	MISS (AU)	MISS (LD)	DIAMETER (miles)	SPEED (km.sec)
2007 AF 2	Jan 21 2007	0.1038	40.4	87	6.55
5011 Ptah	Jan 21 2007	0.1982	77.1	0.6–1.4	6.28
2002 MB26	Jan 23 2007	0.1741	67.7	162–360	7.70
2007 BS2	Jan 24 2007	0.1470	57.2	35–81	2.56
2007 AH12	Jan 25 2007	0.1582	61.6	130–292	4.48
2001 CP36	Jan 27 2007	0.1237	48.2	30–68	7.05
2007 AT2	Jan 30 2007	0.1522	59.2	193–435	4.92
1999 SK10	Jan 30 2007	0.1153	44.9	211–48	4.00
2006 CJ	Jan 31 2007	0.0264	10.3	149–329	13.01

Note:
AU = distance earth to sun (1 AU = 93 million miles)
LD = distance earth to moon (1 LD = 239,000 miles)

And so, in one month, 9 asteroids passed by our neighborhood. The closest miss was the last day of the month, when a building-sized chunk of rock called 2006 CJ whizzed by at 13 miles per second or 46,841 miles an hour, and missed us by a little over 1.9 million miles.

Table derived from NASA data, at http://neo.jpl.nasa.gov/risk/.

The lumpy sphere (NASA has obtained a grainy photograph of the thing, looking like an oversized and rather misshapen walnut) that came to be called 1950 DA is less than a mile in diameter and spins on its axis every 2.1 hours, the second-fastest spin rate ever observed. It was first picked up in 1950, tracked for seventeen days, and then faded from view.

Fifty years later, on New Year's Eve 2000, New Century's Eve and 200 years to the night of the very first discovery of an asteroid, Ceres, Australian astronomers picked up an object they soon identified as the long-lost 1950 DA. It was around 4.85 million miles away, or about 21 times as far as the moon.

Jon Giorgini of NASA's Jet Propulsion Laboratory calculated its orbit and, based on what was known of its velocity and physical characteristics, suggested that there was a remote possibility, maybe 1 in 300, that it would collide with Earth—some 870 years in the future, in March 2880.[7] At the time, 2002, this was the only asteroid whose potential hazard was greater than the general background noise of all asteroids together and didn't seem frightening. It was merely technically interesting.

Apophis, or catalog number 2004 MN4, was a different and more dangerous creature altogether. It was tracked for several days after Christmas 2004 and caused a flurry of alarm to spread through the astrophysical community. That it didn't cause more of a stir was because the Sumatran tsunami had pushed other stories off the news pages. Still, the alarm was there: there was a very good possibility that the asteroid would collide with the Earth, not 800 years from now but in just two decades, on April 13, 2029. Apophis wasn't as large as 1950 DA. It was "only" 1,300 feet in diameter, larger than Tungusta but not a planet buster. Still, it could obliterate, say, Texas or a cluster of European countries.

The NEO that came to be called Apophis was first spotted on June 19, 2004, by a Hawaiian astronomer named David Tholen, using a telescope at the University of Arizona. He took three pictures that night and three the next night, but storm clouds and the moon blocked further observations. Six months later, the same object was spotted again, in Australia, by Gordon Garradd, at the Siding Springs telescope in Coonabarabran, and given its catalog number, Asteroid 2004 MN4. It was given a 1-in-170 chance of hitting the Earth, but as the scientists tracking it refined their calculations the probability went up sharply, and five days later it was standing at 1 in 38, by far the most dangerous space object ever tracked.

Later observations changed observers' minds: Apophis would miss by somewhere between 15,500 and 25,000 miles, still a perilously near thing by cosmic standards.

A close watch was kept on it nevertheless, because there was a suggestion that if by mischance Apophis passed through a small "gravitational keyhole" of about 1,300 feet, its trajectory would change sufficiently to set up a future impact in 2036. As of October 2006, however, the risk was lowered to only 1 in 45,000, and an additional impact probability for 2037 was lowered to a meager 1 in 12.3 million. Even so, a close encounter with the Earth may change its orbit and once again bring it under scrutiny.

So this was no false alarm. As *The Washington Post* writer Guy Gugliotta put it, "It has provided the world with the best evidence yet that a catastrophic encounter with a rogue visitor from space is not only possible but probably inevitable."[8] And the Planetary Society is still offering a $50,000 prize for the best plan to put a tracking device on or near the asteroid, tagging it like an errant goose.

If you followed the technical discussions of Apophis on astrophysical websites during its transit, you would have seen it referred to first as a Torino 1 object. That changed briefly to Torino 2 and, as the hazard assessments climbed sharply, to a Torino 4. It is still the only Near Earth Object ever to have been given a Torino 4 rating.

Which means what?

The Torino Scale, which has been adopted by the International Astrophysical Union, was devised by Richard Binzel of the Massachusetts Institute of Technology, partly to assign a common lexicon to discussions of near Earth asteroids and comets, and partly as a way of calming a media he felt overly prone to hysteria and panic—every time an astronomer mentioned anything "nearby" in space, stories appeared of imminent collisions, environmental collapse, and social panic. "It's very hard," Binzel told the journal *Science* with some asperity, "to communicate extremely low probabilities to the general public." This has partly to do with the immense distances involved. "Near" by cosmic standards can be literally millions of kilometers away. To devise his scale Binzel called on the help of two popular science writers he trusted, Kelly Beatty of *Sky & Telescope* and David Chandler of *The Boston Globe*. "We were acting as sociologists as much as scientists," he said, thereby innocently exhibiting the hard scientist's rather condescending view of his softer-science cousins.

The idea for a kind of Richter scale for asteroids and comets was presented to a meeting of astrophysicists in Torino (Turin) in 1999, hence the name. The most recent version, which rates objects on a scale from 1 to 10, assesses an object's size and speed (and therefore its latent kinetic energy) as well as the possibility of a collision. NASA's Office of Space Science has since adopted the scale, calling it a "major advance in our ability to explain the hazard posed by a particular [object]."[9]

Zero on the Torino Scale indicates that an object has no chance of hitting the Earth (or, if it does hit, that it has zero chance of penetrating the atmosphere, being too small). Ten means certain impact and planetary disaster.

For convenience in this age of color monitors and broadband communications, a color overlay has also been devised. Each color code has an overall meaning:

- White, the No Hazard color, corresponds to 0 on the scale. White is the not-to-worry color.
- Green means Normal, which refers to objects that will get close, with some small but not alarming possibility of a hit. Green corresponds to Category 1 on the scale; most green objects are subsequently pushed into the white column as calculations about their orbits are refined.
- Yellow is the Merits Attention color. These are objects that have higher collision chances than usual, measured as averages over several decades. Yellow means that astronomers should pay close attention. The color corresponds to Categories 2, 3, and 4.
- Orange is getting more dire. It is the Threatening color, corresponding to Categories 5, 6, and 7. Orange means an object that is large enough to cause regional or even global cataclysm, and its chances of colliding are much greater than a century-long norm. If orange shows on the board, astronomers will drop everything to refine its orbit and its hit-rating. So far, nothing has been found close enough to rate an orange.
- Red, for Categories 8, 9, and 10, effectively means game over. It means a certain collision by an object large enough to cause substantial local damage (an 8), regional devastation (a 9), or global catastrophe (a 10).

Tungusta would have been an orange, and if it had been tracked would have shifted rating to a Red 8 at the last moment. Barringer

Crater in Arizona would very likely have been Red 8. Sudbury would have been Red 9 or Red 10, and the dinosaur killer at Chicxulub would have been Red 10. Apophis, briefly a Yellow 4, is the only object, at time of writing, to have shifted beyond the Green.[10]

Until recently, hardly anyone was paying attention to the potential hazards of interplanetary impact. This had something to do with intellectual fashion—catastrophists have been in and out of favor, but, since Lyell and Darwin, mostly out. This wasn't helped much by the writings in the 1950s of a physician named Immanuel Velikovsky, who published a series of books maintaining that planetary impacts (not just asteroids, but whole planets, including Mars and Venus) had occurred within historical and folk memory. His books were best-sellers, partly because his theories seemed plausibly to connect vast ranges of implausible events, and partly because he was widely denounced by orthodox science, which proved to the conspiracy-minded that he must be doing something right, *pace* Galileo. In this scientific climate of ridicule, any real scientist who thought to propose an extraterrestrial origin for any past catastrophe tended to think better of it and keep his thoughts to himself.

A few astronomers went on looking, among them Brian Marsden, who became—and still was in 2007—a clearinghouse of sorts for catalogs of near Earth objects, and David Morrison, an astronomer at NASA's Ames Research Center. "I used to say the total number of people interested in this was no more than one shift at a McDonald's restaurant," Morrison told *The Washington Post* in 2005. "Now it's maybe two shifts."[11]

That all changed after Walter Alvarez persuaded the world that a rogue asteroid hitting the Earth had indeed wiped out the dinosaurs. At the turn of the millennium, the U.S. Congress allocated some funds for research, and a dozen laboratories and observatories started looking. The U.S. government's Spaceguard Survey was assigned to detect and catalog 90 percent of NEOs a half mile and larger by the end of 2008, later extended to 2012. The National Research Center has recommended a search program to locate Earth-crossing asteroids less than a half mile long, focusing on those down to 1,000 feet in size; an internal NASA

NEO Science Definition Team is planning to catalog those as small as 460 feet. Other search programs are being run by the Lincoln Near Earth Asteroid Research program (LINEAR), by Near Earth Asteroid Tracking, by Spacewatch, and by various others, such as the Lowell Observatory Near Earth Object Search (LONEOS), the Catalina Sky Survey, the Japanese Spaceguard Association (JSGA), the Asiago DLR Asteroid Survey (ADAS), and more, including studies by the U.S. Air Force, by the Livermore and Los Alamos weapons labs, by the Russian defense industry, and by the UN.

Perhaps the most significant development is NASA's funding of the JPL (Jet Propulsion Laboratory) Sentry System software, an automated collision monitoring system that continually scans the most current asteroid catalogs, projecting possible collisions a hundred years into the future. (You can see the constantly updated collision risk table at http://neo.jpl.nasa.gov/risk/.) For the paranoid, it is interesting to check in daily to see if new objects have been added and, at the other end, to see existing objects deleted as their trajectories indicate waning risks.

Sentry is designed to detect nearly all potential impacts with probability greater than 10^{-8} (1 in 100 million). It doesn't bother much with lesser probabilities. And it basically ignores all PHAs (Potentially Hazardous Asteroids) that won't get any closer than 0.05 AU (about 4,600,000 miles) or that are smaller than 500 feet. At time of writing there were 842 known PHAs on the list.

As to the damage they'd cause, NASA has given the possibilities a ranking: "Bolides" are rocks up to 5 megatons (5,000 tons)—these would cause great fireworks, but no real damage. Tungusta-class asteroids (up to 15 megatons)—city killers, causing damage similar to a large nuclear bomb. Larger local or regional impacts (up to 10,000 megatons) would destroy an area the equivalent of small country. Global catastrophes (greater than 1 million megatons) would cause global environmental damage and threaten human civilization, killing more than 1.5 billion people. This is the only known natural hazard that could sterilize the planet.

As NASA puts it, deadpan: "This latter is well above thresholds for regulatory action." Meaning: the risk of this happening is low—and what the hell could we do about it anyway?[12]

In 2003 the *Geophysical Journal* published an analysis of what a rock such as the half-mile-sized 1950 AD would do if it plunged into the north Atlantic some 300 miles off the U.S. coast at its given speed, 3,800 miles an hour. The study was conducted by two scientists at the University of California Santa Cruz: Steven Ward, a researcher at the Institute of Geophysics and Planetary Physics at UCSC, and Erik Asphaug, an associate professor of Earth Sciences.

The results were, not to put too fine a point on it, awe-inspiring. The 60,000-megaton blast of the impact vaporizes the asteroid and blows a cavity in the ocean 10.5 miles across and all the way down to the sea floor, which is nearly 2 miles deep at that point. The blast excavates some of the sea floor. Water then rushes back in to fill the cavity, and a ring of waves spreads out in all directions. Two hours after impact, 400-foot waves, the size of a 40-story building, would reach beaches from Nova Scotia to Cape Hatteras; two hours after that, 200-foot waves would hit the entire East Coast and pretty much wipe out most of the Caribbean. Eight hours after impact, the waves reach Europe, where they come ashore at heights of about 30 to 50 feet.

As though that weren't enough, the impact waves would probably destabilize undersea slopes, causing landslides that would trigger secondary tsunamis. If the existing unstable slope in the Canary Islands were to cascade to the seabed, another series of 300-foot tsunamis would batter the American coastline, traveling inland more than 14 miles (such a tsunami would totally submerge Holland and Denmark on the other side of the Atlantic).

At least one such impact hit the shallow waters off America about 35 million years ago. "A crater at Chesapeake Bay was caused by an asteroid or comet traveling at about [70,000 miles] an hour that splashed through several hundred [feet] of water and several thousand [feet] of mud and sediment," the geophysicist Patricia Lockridge wrote in 2002. "It is thought that billions of tons of ocean water were propelled into the air as high as 30 miles and vaporized. Millions of tons of debris and rocks were also ejected into the atmosphere."

Just like the model, the projectile probably incinerated everything along the East Coast, triggered gigantic tsunamis affecting coastal areas on both sides of the Atlantic, and decimated marine life in the surrounding areas.[13]

No evacuation plan could deal with such an impact. If you were in Washington, D.C., say, to be safe you'd have to drive west into the hills at highway speeds for at least two or three hours to get out of range of the tsunami impact and the falling debris. But you'd have, at best, five minutes.

EIGHT

Earthquakes

Which would you rather have, a bursting planet or an earthquake here and there?
—John Joseph Lynch, Archbishop of Toronto, in 1888

The skin of the planet, this thin crust we're standing on, is constantly quivering.

About 8,000 mini-earthquakes take place every day across the planet, nearly 3 million of them a year. Each day, on average, there are 2 quakes greater than magnitude 2; each year 18 major ones; and in most years a really big one, greater than magnitude 8, potentially a city killer.

To the Earth itself, even the major quakes are of no real consequence, no more than a minor irritant, a tiny itch easily scratched, a minuscule adjustment, a truly trivial thing. To those of us living on the skin itself, they are of course of somewhat more moment. Of all the potentially calamitous events that can happen, earthquakes, which make the very ground heave and shake, are the most unsettling. An earthquake can come at any time, at any place, at any strength, with the unpredictability and ferocity of a wrathful god. We have all seen the pictures of survivors picking through the rubble of ruined cities, forlornly looking for the miracle of survival. No wonder they are terrifying.

We were in Santa Barbara, 50 miles or so north of Los Angeles, when the Northridge earthquake hit at 4:30 on a January morning in 1994; we'd stayed overnight at a B&B near the center of town, wanting to get out early the next morning for the Pacific Coast Highway and then San Francisco. Well, 4:30 was early, but it wasn't an alarm clock that woke us. Instead, we woke to the rattling of the bedstead against the wall, then the tinkling of breaking glass, and suddenly the whole bed was shaking, like a car going at high speed over a rutted road. Someone the previous night had warned us that the beds in the place were noisy—"They'll hear everything you get up to," he'd said—but of course he hadn't been talking about the Earth moving in such a literal way.

The shaking didn't last more than ten or fifteen seconds. When it was done, we stumbled out into the hall. The power was already off, and it was still dark outside.

The door across the hall opened and a figure emerged. "Stay still," a voice said. "Don't move. There'll be aftershocks coming."

It was a U.S. Marine, traveling with his girlfriend. What a relief! He came complete with flashlight, emergency radio, every possible Boy Scout Be Prepared device, and knew at once to go to the basement to check for ruptured gas lines, taking charge in a very efficient military way. We just waited, and sure enough a minute later the house shook again, and somewhere a mirror fell over, or off a wall, with a crash, and then it was silent.

After a while we went back to our room and sat on the bed, waiting for dawn. The house seemed to have come through without any damage. It was very quiet outside. No traffic sounds, no distant sirens, nothing. Then there was a squeaking sound, followed by a thud, and I looked out of the window. A boy on his bicycle had just delivered the morning paper, flinging it onto the porch without stopping. My wife rolled her eyes. Great, she said, that's the press, up to the minute, yesterday's news delivered today. . . .

We went out to the car and checked the gas. Less than half a tank, but probably enough to get us out of the quake zone to a gas station with power for the pumps, provided the highways were intact and no overpass had fallen to block the roads. We checked the Marine's radio to get the news. The morning DJs were babbling about the quake, but no one really had any information yet, except that it was strong, an elevated expressway had collapsed somewhere, the epicenter was somewhere around Northridge, some 20 miles northwest of downtown Los Angeles, not that far from where we were.

Later we called friends who had been in a hotel in Santa Monica, near the beach. Somehow the hotel had got an indicator that something was going to happen, because a disembodied voice emanating from the ceiling had ordered everyone out of the building. "Everything was in seven languages, so it took forever to say anything," the male of the couple said. "We were getting out of the room, but she"—no doubt gesturing to his wife—"insisted on doing the cosmetics thing and she took five minutes in the bathroom, and then we went into the corridor and . . ." Here he stopped for a minute. ". . . and we saw the hotel floor . . . *rippling* . . . toward us. Like a small wave coming in off the ocean, I mean a solid floor *rolling* toward us . . ." They got out in a hurry and joined the thousands of pajama-clad residents of the area huddling together on the beach, and just waited there.

They, and we, were lucky. Almost sixty people died in those few minutes. Sections of major freeways had indeed collapsed, a brand-new 2,500-car parking garage fell in on itself, office buildings tumbled, landslides occurred in the Santa Susana and Santa Monica mountains. A motorcycle cop racing to see if he could help fell into the void where an expressway had been minutes before and was hurled to his death.

The quake had been the strongest recorded in a built-up area of California, at 6.7 on the Richter scale; the aftershock that followed was 5.6. More than a thousand smaller aftershocks were measured.

It wasn't the San Andreas fault of dread expectation, though it was nearby. The slippage occurred on a blind thrust fault (one that was fully underground and not properly mapped) named the Northridge Thrust (or Pico Thrust), 11 miles beneath the surface; several other small faults suffered minor ruptures in the aftershocks.

W e and our friends were lucky in another way too. This was a pretty trivial earthquake, as these things are measured. There have been many worse, in America and around the world.

The strongest earthquake yet recorded was in Chile in 1960, thought to be 9.5 on the Richter scale. (Bear in mind that the scale is logarithmic, and that "one degree up is not one degree quakier," as Richard Fortey put it.[1] A 7.3 quake releases 50 times more energy than a 6.3, and is 2,500 times more powerful than a 5.3.) Three of the world's largest twentieth-century quakes happened in Alaska, in 1957, 1964, and 1965, including the second-largest ever recorded, at Prince William Sound 81 miles east of Anchorage, on Good Friday in 1964, at 9.2 Richter. It permanently rearranged 77,000 square miles of landscape (the surface rose by as much as 33 feet in places, dropped 6.5 feet elsewhere) and caused a tsunami 49 feet high that traveled more than 8,100 miles at 280 miles an hour.

The deadliest quake of all time was of unknown magnitude— Richter hadn't been born in 1550—in central Shaanxi province in north central China. Hundreds of thousands of people made their homes in caves dug into the soft loess hillsides that characterize the region. Most of these caves collapsed in the quake, burying and killing more than 800,000 people.

Earthquake death tolls are much worse than those of, say, volcanoes, because volcanoes often give advance warning of eruptions, whereas earthquakes come in entirely unpredictable places and at unpredictable times. The accumulated death toll makes grim reading: San Francisco, 1906: a low death toll of perhaps 3,000, but 29,000 homes destroyed by quake and fire, and 225,000 people homeless; Messina, Italy, 1908: 70,000 to 100,000 dead; Tokyo-Yokohama 1923: 143,000 dead; Gansu, China, 1932: 70,000 dead; Mexico City, 1965: 9,000 dead; Peru 1970, setting off an avalanche that carried 80 million tons of rock, mud, and snow at more than 100 miles an hour, killing 70,000 people and leaving 600,000 homeless; Tangshan, China, 1976: 255,000 dead; Kashmir, 2005: 87,350 dead, 3 million homeless.

Probably the worst urban earthquake calamity was the tremor and resulting tsunamis that struck the Portuguese capital of Lisbon on All

Saints Day, in 1755. Only about half of the city's 275,000 people survived. All those saints didn't help. Not a single one of the forty churches in Lisbon survived intact.

The great organ was playing in the cathedral at 9:40 in the morning when the majestic sounds of the Kyrie eleison, the Mercy of God, were drowned by a thunderous roar coming from the floor, and the massive stone edifice swayed and groaned as though Satan himself were shrugging his shoulders. Those near the doors scurried into the street, to see it rippling and bucking in a series of great waves, then the whole city swayed and the cathedral they had just left collapsed in a heap of rubble, crushing those still inside. Lisbon was made of stone, and all the stone buildings collapsed.

Many of the survivors fled to the shore, where there was open space free of buildings. But it did them no good; the earthquake had been offshore, a great shelf had collapsed, and tsunamis were sent racing landward.

The Portuguese capital is not on the ocean, but 7 miles up the Tagus River, which at that point is about 2 miles wide. The first of the three great tsunamis, as high as a four-story building, poured upstream about forty minutes after the quake, scouring both shores of buildings, tearing ships off their moorings, and smashing the resulting heap of splintered wood and rubble into the wharves where the fleeing people had assembled. No one could have survived. It was followed by two more waves.

The five hundred aftershocks that followed were quite minor, but it no longer mattered: fire was destroying what the earthquake and the waves had not; this disaster was followed soon enough by another Horseman of the Apocalypse, famine.

The king of Portugal, José I, and his family escaped. He'd left the city after attending a dawn mass, mostly at the urging of one of the princesses, who wanted to spend the holiday in the countryside. He returned to the city as soon as the news reached him and set about organizing a system of relief. Martial law was declared, and looters executed. But the king never really recovered and developed a morbid fear of buildings. A new court was set up in a complex of tents and pavilions in the hills of Ajuda, then outside the city but now a suburb. A proper palace wasn't built until after his death, and a new royal home was erected by his daughter, Queen Maria I. José's chief minister, Pombal, was given authority over the reconstruction, and in the absence of any real royal authority he used his new position to set up a two-decade career as dictator.

The initial earthquake, with its epicenter about 125 miles southwest of Cape St. Vincent, was probably between 8.75 and 9 Richter, making it one of the two or three strongest in the last 250 years. It devastated not just Lisbon but many other coastal cities. The destruction through the Algarve was immense. At the Spanish port of Cadiz, 217 miles to the south, the first tsunami was measured at 18 feet; other reports indicate that the maximum runup was as much as 66 feet, as high as it was in Lisbon itself. Shock waves were felt as far away as Finland. Tsunamis up to 66 feet struck the North African coast and swept across to Martinique and Barbados. A 10-foot tsunami hit the southern English coast and Galway in Ireland.

The quake was widely discussed in Europe. It was the first earthquake to receive any real scientific study (and the first, also, to raise the notion that some animals had sensed danger and had fled to higher ground before the giant waves arrived). It also precipitated a brief fashion for theodicy, the always-vain attempt to reconcile the existence of suffering innocents with the notion of a benevolent god. Enlightenment philosophers seized the moment to further secularize public discourse, preparing the way for, among other things, the advent of geological realism and thus, indirectly, Darwin.

It's not easy to measure earthquakes. Several scales of varying utility have been developed, and several others were once used but have since fallen into disuse. They measure both intensity, which is based on the observed effects of an earthquake, and magnitude, which is a measure of the seismic energy released.

Intensity measurements, obviously enough, are more subjective than magnitude ones. The first known one was a simple from-bad-to-worse scale devised by Italy's Domenico Pignataro in the 1780s; but the first of any real use was arranged by P.N.G. Egen in 1828, quickly followed by a raft of others, including the Rossi-Forel scale and many more. The Medvedev-Sponheuer-Karnik Earthquake Intensity Scale, for pretty obvious reasons of economy usually called the MSK, was until recently widely used in Europe and Asia and is still current in Russia. It uses a scale from 1 to 12 and uses parameters such as how the quake is experienced by people, damage to buildings and structures, and impact on the landscape. It has mostly been abandoned, supplanted in 1988 by the EMS, or European Macroseismic Scale, which ranges from "Imperceptible, Effects Not Felt" through "Strong, Causing Fright" all the way to "Very Catastrophic, Upheaval of Landscape, Tsunamis, Structures Destroyed."

The most widely used intensity scale is now the Mercalli, or Modified Mercalli (MM) scale, which also measures increasing levels of intensity from imperceptible shaking to catastrophic destruction.

The best known of the magnitude measuring devices is the Richter scale, which is now mostly used by journalists. (Where scientists do use it, it is not called the Richter scale but the M_L scale, for local magnitude scale.) It was developed at California Institute of Technology in 1935 by Charles F. Richter. It was, as already explained, a logarithmic scale of measuring energy and ranges from 1, which is barely noticeable, to 9, the strongest earthquake yet recorded on Earth. Scientists stopped using it because although it was precise enough for the lower orders of quakes, it was not as good at measuring strong earthquakes. The most popular measurements are now those of "surface-wave magnitude," the seismic waves crackling around the Earth's surface, and "moment magnitude," which is based on the size of the fault on which an earthquake occurs and the amount the Earth slips.

Earthquakes happen when stresses built up by plate movements in a geological fault cause an abrupt slip. As the engineer Ernest Zebrowski describes it, the kinetic energy thus released spreads outward from the source in a series of more or less spherical wave fronts. Deep down, these seismic waves bend and deflect where the crust and the mantle meet and, deeper down, at the boundary between the mantle and the outer core. At the upper surface, which interests us rather more directly, they transfer some of their energy to the seas (causing tsunamis) and the atmosphere (the origin of the "rumbling" sound often associated with quakes). The remaining energy, usually a rather substantial portion of it, radiates outward along the crust like a series of ripples in a pond.[2] It is one of the eeriest feelings of all to see the apparently solid Earth rippling, for all the world as though it were water. The nature of the terrain can amplify the effect. Where the soil is deep and wet, or where subdivisions are built on landfills, as in parts of the Embarcadero in San Francisco, these seismic waves slow down in the same way that tsunamis do when they enter shallow water: they slow, but grow in amplitude and destructiveness. In 1985, for example, a destructive quake hit Mexico City, but most of the damage was confined to a single 10-square-mile section of the city that had been built on an old lake bed.

Earthquakes have another, more hidden, effect. Studies after the Andaman Sumatran earthquake that set off the great tsunami of 2004 suggest that quakes can somehow activate volcanoes in an arc around the epicenter—even dormant, or rarely erupting, volcanoes.

It is obvious that most earthquakes happen at tectonic plate boundaries—at faults, which are really fractures in the Earth's crust caused by rocks cracking under the inexorable stresses set up by plate movement. Indeed, it is the fact of earthquakes that have facilitated the mapping of the plates themselves. But earthquakes can happen anywhere, not just on plate boundaries. Intraplate quakes can represent ancient stresses that built up when the landscape was being created eons ago. Quakes in those regions can occur on now-invisible "faults" that have lain dormant for half a million years or more and whose surface has long ago eroded, making them to all intents and purposes invisible.

There are different kinds of plate boundaries, and the stresses they cause result in different kinds of earthquakes.

Some of the plates are moving away from each other, causing spreading ridges; these are called extensional boundaries. The mid-ocean ridges are extensional. So is, for example, the Juan de Fuca ridge off Vancouver Island. Extensionals usually cause shallow, almost surface, earthquakes that are usually (but again not always) lower than magnitude 8.

Other plates are moving past each other or may be changing from sliding to either extensional or its opposite, compressional boundaries. The San Andreas fault is one of these. In fairly recent geological time, perhaps twenty-five million years, a convection eddy deep underground changed the direction of the Pacific plate's movement, from what was thought to be eastward to the current northwest movement, transforming what would have been a compressional fault to a "transform" one, where one plate slips past another. Earthquakes in these faults are caused by sections of one plate snagging on the other as they slide past each other, causing the occasional abrupt lurching motion. Transform faults don't usually cause much uplift or subsidence, but mostly lateral movement. You can see this clearly in Hollister, California, which lies pretty much directly on the fault, where fences, roads, and occasionally even parts of houses have shifted in a zigzag motion in relatively short periods. Transform earthquakes tend also to be shallow, no more than 15.5 miles or so deep, and, though destructive enough, have not produced an earthquake greater than magnitude 8.5.

The largest earthquakes are compressional ones, in which plates collide, causing subduction, forcing one plate underneath another, making one sink back into the mantle, the other rear upward in massive crumpling. A single earthquake can thrust thousands of square miles of land many meters into the air. These earthquakes may also be near-surface, but are typically very deep, as much as four hundred miles down, because the subducting plate, not yet warmed by the molten mantle, is still brittle enough to snap. Compressional quakes are the largest known, magnitude 9 and possibly greater. Both Alaska and Chile are at compressional boundaries. So are Anatolia, Turkey, and the Aegean. So are Kashmir and the Tibetan plateau. So are Java and its neighbors. And Japan. There are many perilous places on this planet.

Plate movements generate more than just the abrupt slips we call earthquakes. Deeper, slower forces are also at work.

For example, early in 2007 seismologists noticed an unusual traveling tremor deep inside the crust on the west coast of North

America. It was called a slow-frequency seismic event, or a slow-slip event. It began in the Puget Sound area off the Washington and British Columbia coasts and traveled northward along the Juan de Fuca plate at a rate of about 6 miles a day, passing slowly underneath Vancouver Island.

This was worrying—slow-slip events are not just background noise, but represent slower and deeper earthquakes. Was this causing further stress or alleviating it? Herb Dragert, a seismologist with the Geological Survey of Canada in the British Columbia capital, Victoria, explained it this way:

> The most awesome tectonic characteristic of the world's subduction zones is the repeated occurrence of great thrust earthquakes. Although the average recurrence intervals for these devastating temblors are directly related to plate convergence rates, the nature of the stress buildup that culminates in these large fault ruptures is not well known. . . . [But] scientists agree that it is the inexorable aseismic plate motions (that is, motion without earthquakes) at depth that increase stress on the overlying portions of the fault, [and] that will ultimately be relieved by an earthquake. Until recently, it was assumed that this deep aseismic slip or ductile shearing between plates occurred at a uniform speed. But the observation of small (less than 5 mm) systematic surface displacements over a period of several weeks led to the discovery of a silent slip event on the Cascadia Subduction Zone [Washington and British Columbia]. . . .[3]

The rock itself is changing under the intense pressure, Dragert said. "It's not correct to say the rock is melting, but the mineralogy is changing. There are fluids and volatiles trapped in the rock and they are being expelled. It's a kind of dehydration, you could say it's being cooked out." At 15 to 30 miles below the Earth's surface, where temperatures reach 1022 to 1112 degrees Fahrenheit, these volatiles act as a sort of lubricant to facilitate slippage.

One of Dragert's colleagues, Tim Melbourne, put it this way in an interview with the *Victoria News* in February: "The reason it's kind of spooky is we now know that stuff down there is on the move," he said. "But there's a big difference between an earthquake and these tremors. With an earthquake, the fault breaks and slips very quickly." Rather than releasing pent-up seismic energy, Dragert said, the slow slips actually increase tension, albeit in tiny increments, increasing the likelihood of a major quake.

"We know it adds to the stress, but we don't know what the breaking point is," he said. "The tension is creeping up in discrete steps and eventually one of these discrete steps will trigger a major quake. We just don't know when."[4]

The size, and nature, of an earthquake or seismic rupture is controlled by the nature of the underlying fault. But this can also change over time, for faults can spread. Also, the more a fault slips over its lifetime, the smoother it becomes, and smoother faults make larger earthquakes. The size of the fault is critical; an earthquake will continue along a fault and continue to propagate, unless it reaches a barrier of some sort. Until recently, no one knew what this barrier could be, but it had to be a structure where the stresses are low enough to stop rupture. As it turns out, an earthquake can't spread if it comes to a discontinuity, or gap in the fault, wider than about 2 miles. A dramatic proof was the Christmas 2004 Sumatra-Andaman earthquake, the first major quake to happen in the modern era of space-based geodesy and broadband seismology, which produces massive data sets for analysis—more than 300 observation posts, in this case. The observations showed that the fault slip had been massive—a rupture of the Sunda "subduction megathrust" caused a lurch of up to 66 feet at depths shallower than 12 miles. The rupture was 932 miles long and almost 93 wide, but it came to an abrupt end to the south of the uplift, because of just such a discontinuity.[5]

Some earthquake-prone areas seem to store up seismic energy instead of dissipating it. Scientists had often held that earthquakes were caused by stress buildup, and this is still generally thought to be true. But the corollary, which they called the seismic gap theory, was that if you experienced a major quake in any one area, it should be quake-free for a long time afterwards. The reality is somewhat more dire. The seismologist Roger Bilham and his colleagues conducted a study of recent Himalayan earthquakes, including the devastating Kashmiri 7.5 quake of 2005, and concluded that "two thirds of the Himalayas are ready for an earthquake of magnitude 8 or greater . . . [and] some are long overdue." The research showed that the entire Himalayan plateau is much more elastic than previously believed and can return to its original shape after temporary stresses. But this does not mean, as a commonsensical interpretation would suggest, that it should therefore not rupture as easily. On the contrary, it means that a short rupture doesn't drain the system's energy, and that a large earthquake can be followed by another in short order. "It takes megaquakes to drain the system. We've had four substantial quakes in the past 200 years, but they haven't been able to release much of this cumulative strain."

And then, abandoning the measured tone of scientific discourse, they added this: "The countries bordering the Himalayas are in terrible trouble. We estimate that one million people could die in a single event unless city structures are made earthquake resistant."[6]

In October 2007, after a series of powerful quakes rocked Indonesia, geologists were still warning that the energy had not been squeezed from the system and "the big one" was still to come, a deadly 9.0 quake that could set off another massive tsunami.

Istanbul and all of Anatolian Turkey have been on high seismic alert for many centuries, courtesy of the North Anatolian Fault (NAF), actually a network of smaller faults that underlie the Sea of Marmara, among other places, on the boundary between the Eurasian and Anatolian plates. The NAF has caused at least a dozen major earthquakes since 1939, the most recent the catastrophic quakes of August and November 1999, the first of which produced a 13-foot tsunami in the Gulf of Izmit; more than seventeen thousand people died.

Istanbul itself has been shaken five times in the past five hundred years; there are suggestions that no fewer than six hundred quakes have rocked the region in the Christian era. The fault is plausibly the source for the tale of the Biblical flood (see Chapter 12). There have also been suggestions that Homer's Troy was destroyed in an earthquake;"written records of great past earthquakes extend back for two millennia, and reveal a cluster of devastating earthquakes in the Early Byzantine period from the fourth to the sixth century (the 'Early Byzantine Tectonic Paroxysm'),"[7] and many of the major cities of the region, such as Aphrodisias, Ephesus, Pergamum, and Smyrna, were severely damaged. Some writers have suggested that the Trojan horse was actually an offering by the besiegers to the gods for wrecking Troy without human help.

Tokyo is an accident—a very bad accident—waiting to happen. And since the area is one of the most intensely monitored and studied seismically active areas on the planet, scientists are closing in on a still fairly rough but plausible estimate of when another major quake will hit: a greater than 30 percent probability of a major quake in the next thirty years.

This is a very rough seismic neighborhood. Tokyo is not a simple boundary between plates, as parts of California are. Instead, as already suggested, a joint American and Japanese survey[8] has found that the underlying geology is "magnificently complex"; it seems that a massive dislodged block of the Pacific plate is jammed between three plates—its parent the Pacific plate, the Philippine sea plate, and the Eurasian plate, and this jam could, and probably will, become unraveled sometime soon. To make things worse, both the cities of Tokyo and Yokohama are located on alluvium, or soft river delta deposits, which means that a major quake would have results similar to those of Mexico City and the Embarcadero. Tokyo long ago abandoned its traditional wood and rice paper construction and is as steel-and-glass-modern as any other capital. The city and its surrounding region are home to a quarter of the Japanese population, maybe 31 or 32 million people; a Japanese government estimate is that a really severe earthquake (such as those that struck Tokyo in 1703, 1855, and 1923) would kill a million people or more and cause damage greater than the country's annual budget; the city is less than one-quarter insured.

The geologic record shows that there have been 17 quakes of magnitude 8 or greater in the past 7,000 years. The dense modern seismic network that surrounds the capital has detected no fewer than 300,000 earthquakes in the last 30 years, all of them, so far, tiny.

The earthquake that destroyed the Kanto plain (Tokyo, Yokohama, and the surrounding areas) on September 1, 1923, was one of the worst in human history. Somewhere around 140,000 people perished in the quake and the resulting fires. The earthquake struck just before noon, when thousands of charcoal braziers and gas ranges were lit up to prepare the noon meal. The epicenter was in Sagami Bay, just southwest of Tokyo, in close proximity to densely populated regions containing more than 2 million people. Within minutes, much of Tokyo and Yokohama lay in ruins; destruction reached up into the popular tourist resorts in the Hakone mountains and to the busy shipping lanes of Yokohama Bay. Flammable materials in industrial plants fed the fires; at least one munitions factory blew up. Water mains were severed and there was nothing to combat the fires. At one point, 88 fires were raging in Yokohama alone.

Some 30,000 of the dead were people who had fled to an open area in the Honjo and Fukagawa districts, but they were engulfed by a fire-generated cyclone and were incinerated on the spot.

A few days later rumors swept the capital that "foreigners," unspecified, were taking the opportunity in the chaos to annex Japan; these were hastily denied by the government, but sporadic attacks on Koreans were reported.

That western America is earthquake prone must be one of the best-known geological facts of all time. The San Andreas is the most famous of literally dozens of faults that crack California, Oregon, Washington, British Columbia, and Alaska; when "the Big One" does arrive, people will be shocked but not at all surprised. Still, westerners generally remain untroubled by possible disaster. A while ago I spent some time at a winery in the Gavilan mountains of California's Central Coast, an austere but beautiful spot on a steep hillside, at the bottom of which slope was the San Andreas Fault itself; if the Big One did indeed come, the winery and its precious contents of high-end Pinot Noir and Chardonnay would be exceptionally vulnerable. I remember asking the owner, Josh Jensen, if the notion ever bothered him, but he merely gave me that puzzled look that all Californians adopt at the question and never bothered to answer.

O f course, the most famous earthquake in American history was the one that leveled San Francisco in April 1906, the subject of dozens of books, articles, and movies (two recent books have been written about the quake: Philip Fradkin's *The Great Earthquake and Firestorms of 1906* and Simon Winchester's *A Crack in the Edge of the World*). Both books point out that 1906 was, geologically speaking, an annus horribilis, to quote England's Queen Elizabeth (she was speaking of family and domestic matters, not earthquakes, though she could have been excused for conflating the two at times). It began in the last day of January, with a powerful quake (possibly the largest in historical times) off Ecuador, devastating scores of towns, its effects reaching as far as Hawaii. Sixteen days later there was another quake, on the island of St. Lucia in the Caribbean, between 7 and 8 on the Medvedev-Sponheuer-Karnik Earthquake Intensity scale. Five days later another quake shook Shemakha in the Caucasus Mountains, and four weeks after that a huge quake on Formosa (March 17). And then Vesuvius erupted, killing 150 people. Then in April the San Francisco disaster. Before the year was done 20,000 Chileans would die when an enormous quake, 8.3 magnitude at least, struck Valparaiso in mid-August, wiping it out. Much the same thing seems to have happened in 2004—Iran first, then the Sumatran quake, followed by a series of quakes and tremors for the whole twelve months. Were these all connected? As Simon Winchester put it, after recounting all these events, are there indeed bad years in seismic activity? "Thus far there is no firm evidence—only the numbers, and the anecdotes, that show incontrovertibly that some years are seismically very much more dangerous than others."[9]

San Francisco began to shake at 5:12 A.M. on Wednesday, April 18; it was probably about magnitude 7.9. Buildings shook and fell, water pipes fractured, gas mains exploded, electrical lines snapped and arced, and a great firestorm was set off that engulfed the city. The firefighters were helpless without water and took to dynamiting buildings in the way of the fire to stay its course, almost always making things worse.

Perhaps 3,000 people died in the quake and the fires, with more than three times that number injured. Nearly 28,000 buildings were destroyed. Typhoid followed, as it usually does when the clean water supply is contaminated, and a quarter of a million refugees fled, ferried free by the railways to places like Portland, Oregon, and even Chicago far away to the east.

A ll of this is well enough known. Somewhat less known is that the greatest earthquake in America since the states united was not in the west at all, but in the middle of the continental plate, in the Mississippi valley, near a place called New Madrid, Missouri.

The four hundred residents of the little town of New Madrid were wakened early in the morning of December 16, 1811, by a violent shaking and a tremendous roar. Afterwards, survivors described massive cracks in the Earth's surface, the ground rolling in huge waves, fields and whole farms rising and falling. The crew of the steamboat *New Orleans*, the first steamboat on the Mississippi, then on her maiden voyage, had moored their vessel to an island the night before and woke to find the island entirely gone.

It was only the beginning of a winter of terror. Two other major shocks happened on January 23, and then again on February 7; the February shock, now estimated to be around magnitude 8, destroyed what remained of the town and was felt over an area of 96,313 square miles; church bells rang spontaneously in Boston, 994 miles away, and in Charleston, South Carolina; the tremors were felt in New Orleans and Toronto. So violent was the subsidence that the Mississippi River actually ran backwards for a period, streaming upward toward Minnesota; afterwards it changed its course, creating what is now the Kentucky Bend. New lakes were formed, including Reelfoot Lake in Tennessee. Sandblows erupted in fields, engulfing houses and barns; fissures opened and closed, oozing mud and noxious gases; the bluffs around Memphis collapsed. For a nightmarish five months, the ground shook and trembled; more than 1,850 separate tremors were recorded before it was over. Two years later (Hurricane Katrina wasn't the only disaster to cause a tardy political response) the territorial governor of Missouri asked for federal relief for the inhabitants of New Madrid County, the first such disaster relief request in U.S. history.

Was this a complete fluke? Not at all: the Mississippi Basin is the most seismologically active region of America outside the far west. At least three faults are known to exist beneath Missouri. It is even possible that the North American plate is beginning to crack into two, and that in times to come you will need to take a ferry across the Mississippi Sea to get from one side of Missouri to the other. The seismograph in New Madrid virtually never stops quivering. The needles "wave furiously all the time, like antennae of some excitable insects," in Simon Winchester's words.[10] Strong earthquakes in the central Mississippi Valley, in what is now known as the New Madrid Seismic Zone, are not freak events but have happened repeatedly in the geologic past; the geology of the region means that a severe earthquake in the center will affect a much greater area than a simi-lar one in the west; the San Francisco quake was felt no farther than Nevada, 350 miles away. A report from so sober a body as the U.S. Geological Service (USGS), with the heavyweight backing of the U.S. earthquake consortium (a partnership of the federal government and eight states in the central United States), and FEMA, the Federal Emergency Management Agency, trying harder since New Orleans, has said this:

> Strong earthquakes in the New Madrid seismic zone are certain to occur in the future. Earthquakes of moderate magnitude occur much more frequently than powerful earthquakes of magnitude 8 to 9; the probability of a moderate earthquake occurring in the New Madrid seismic zone in the near future is high; such a quake could affect cities all the way to Chicago. Scientists estimate that the probability of a magnitude 6 to 7 earthquake occurring in this seismic zone within the next 50 years is higher than 90 per cent, 63 per cent in the next 15 years. Such an earthquake could hit the Mississippi Valley at any time.

In 2005 seismologists detected a five-year uptick in seismic activity in the region.

In 1811 the central Mississippi region was only sparsely populated. Now, the region has cities like St. Louis and Memphis, home
to millions of people, and with hardly any buildings designed to
withstand shaking. The New Madrid earthquake of 1811 caused
severe landslides on the banks of the river; the USGS points out that
modern Memphis is built on just the sort of bluff that the earthquake
demolished.

No one is doing anything to prepare for any such disaster. When
it happens, it will come as a terrible surprise.

NINE

Volcanoes

By turns hot embers from her entrails fly,
And flakes of mounting flames, that lick the sky.
—Virgil, Book 3, *The Aeneid*

You can peer into the cone of a volcano in several parts of this planet, if you're careful and don't mind a few risks. That irascible volcano Vesuvius is a good candidate, if you watch out for the traffic and dodge the tour buses up near the summit. The gas that seeps out is sulfurous and choking, like a really bad smog day in Los Angeles, and though it's hard to get close enough to actually see anything interesting, you can hear faint rumblings and occasional clatterings not sounding any more ominous than a massive belly growling after too much rich food. Sometimes, if the wind is right, the chattering of the tourists or the sharp buzzing of a moped down in the valley is louder than the mountain. But that's because Vesuvius is now in one of its unpredictable quiet periods.

It wasn't always so. This most famous of all volcanoes (okay, Krakatoa might be a rival in popular affection) has been busy all through historical times. Most famously, of course, it suddenly exploded on August 24, 79 AD, destroying the Roman cities of Pompeii and Herculaneum. Vesuvius had been considered extinct up to that time; its surface and even crater were covered with lush vegetation. In just a few hours, hot volcanic ash and dust buried the two cities so thoroughly that their ruins were not uncovered for nearly 1,700 years.

This eruption wasn't the first, or the greatest, that Vesuvius has produced. In 3780 BC or so, a massive blast buried much of the Campanian region and its Bronze Age inhabitants. And since the Pompeii eruption, the mountain has been keeping people awake, with numerous minor eruptions and many big ones—in 1631, 1717, 1794, 1872, 1906, and again in 1944, catching the attention of troops then rumbling through Italy in the Italian campaign of World War II. The 1631 eruption was bad, killing twice as many people as the Roman one. The rich volcanic soil and consequent growth that had once again covered the mountain had encouraged human settlement around the cone, and the population had no warning when the mountain suddenly blew on the morning of December 16. The explosion was large enough that within a day, the ash had reached Istanbul.

Pompeii was buried and its two thousand victims forgotten, but its passing did not go unrecorded. That remarkable polymath Pliny the Younger, friend to emperors, survivor of the murderous Domitian, and himself a consul, was in Naples when the mountain blew up. Pliny was also an acute observer of nature. Pliny the Elder, uncle to Pliny the Younger, was killed in the eruption, and the younger Pliny was asked by the historian Tacitus to describe what happened. His letter to the historian is sometimes said to be the beginning of the science of volcanology. He described the eruption thus:

> [B]etween 2 and 3 in the afternoon my mother [noticed] a cloud of unusual size and appearance. . . . The cloud was rising from a mountain—at such a distance we couldn't tell which, but afterwards learned that it was Vesuvius. I can best describe its shape by likening it to a pine tree. It rose into the sky on a very long trunk from which spread some branches. I imagine it had been raised by a sudden blast, which

then weakened, leaving the cloud unsupported so that its own weight caused it to spread sideways. Some of the cloud was white, in other parts there were dark patches of dirt and ash.

Pliny's uncle, the Roman fleet commander in the region, readied a boat to see what he could find out, and to rescue those whom he could. It was an ill-fated expedition; the winds drove them to shore, and they couldn't leave.

Ash was falling onto the ships now, darker and denser the closer they went. Now it was bits of pumice, and rocks that were blackened and burned and shattered by the fire. Now the sea is shoal; debris from the mountain blocks the shore. The wind carried my uncle right in. . . . In order to lessen . . . fear by showing his own unconcern, he asked to be taken to the baths. He bathed and dined, carefree or at least appearing so (which is equally impressive). Meanwhile, broad sheets of flame were lighting up many parts of Vesuvius; their light and brightness were the more vivid for the darkness of the night. Then he rested, and gave every indication of actually sleeping; people who passed by his door heard his snores, which were rather resonant since he was a heavy man. The ground outside his room rose so high with the mixture of ash and stones that if he had spent any more time there escape would have been impossible. He got up and came out, restoring himself to [his friend] Pomponianus and the others who had been unable to sleep. They discussed what to do, whether to remain under cover or to try the open air. The buildings were being rocked by a series of strong tremors, and appeared to have come loose from their foundations and to be sliding this way and that. Outside, however, there was danger from the rocks that were coming down, light and fire-consumed as these bits of pumice were. Weighing the relative dangers they chose the outdoors; in my uncle's case it was a rational decision, others just chose the alternative that frightened them the least.

They tied pillows on top of their heads as protection against the shower of rock. It was daylight elsewhere in the world, but there the darkness was darker and thicker than any night. But they had torches and other lights. They decided to go down to the shore, to see from close up if anything was possible by sea. But it remained as rough and uncooperative as before. Resting in the shade of a sail he drank once or twice from the cold water he had asked for. Then came a smell of sulfur, announcing the flames, and the flames themselves, sending others into flight but reviving him. Supported by two small slaves he stood up, and immediately collapsed. As I understand it, his breathing was obstructed by the dust-laden air, and his innards, which were never strong and often blocked or upset, simply shut down. When daylight came again 2 days after he died, his body was found untouched, unharmed, in the clothing that he had had on. . . .

Volcanoes are the most obvious, and most flamboyant, manifestation of the constant turmoil of the witches' cauldron beneath our feet. They're called volcanoes after the island of Vulcano, in the Tyrrhenian Sea north of Sicily, the southernmost of the Aeolian Islands, near Stromboli, a volcano that has itself been active, on and off, for about 135,000 years, give or take a decade or two—the latest eruption as recently as 1898–1900. In Roman times, it was a given that Vulcano was the chimney of Vulcan's forge, Vulcan being the blacksmith of the Roman gods. The hot lava fragments and clouds of dust erupting from Vulcano came from the forge itself as Vulcan beat out thunderbolts for Jupiter and weapons for Mars. We are now sure, as sure as the Romans were of their theory, that volcanoes are either at tectonic plate boundaries or, not infrequently, over "hot spots," places that are underlain by anomalous hot plumes of mantle dipping down to 125 miles beneath the surface, representing a weakness in the underlying mantle. Around thirty of these superplume hot spots exist around the globe; sometimes they pour out streams of molten lava of the sort that created the Deccan Traps in India. Hot spots are stationary, and the crust slides across their tops. Yellowstone National Park is over such a hot spot now; it was once in Idaho and will sometime in the future move to Wyoming. Other hotspots include the Galapagos Islands and Hawaii.

At least 1,500 volcanoes are active around the world, and somewhere between 10 and 20 are erupting on any one day. But those are only the ones on land; there may be many thousands more under the ocean, only rarely coming to human attention, such as when they appear as new islands, as Surtsey did off Iceland when it popped from the sea in a fire-and-brimstone birth in 1963. Volcanic periods, the average times between eruptions, range from months (Kilauea in Hawaii) to 100,000 years or much more (Yellowstone), though a few volcanoes are more or less constantly erupting, as is Stromboli, an island in the Mediterranean, which has been boiling over for at least 2,000 years—it is sometimes known to sailors as "the Mediterranean's lighthouse." But the gross number of known volcanoes is going up all the time, as more and more potentials are discovered—bear in mind that some of the most fearsome recent eruptions, such as Pinatubo in the Philippines in 1991, El Chichon in Mexico in 1982, and Arenal in Costa Rica in 1968, happened at mountains that were not even recognized as being volcanoes. However, certain "volcanic fields" may contain hundreds of small volcanoes all related to the same magma chamber and therefore should rather be counted as a single entity.

The biggest known volcano is Mauna Loa, in Hawaii, which rears off of the sea floor to 2.4 miles above sea level, a total of 5.5 miles, about the size of Everest. One of the oldest is Etna, a relative pussycat of a volcano that is thought to be more than 350,000 years old. Most are much younger than that—not much more than 100,000 years old.

The most active volcanic belt on Earth is the Ring of Fire, which as mentioned in Chapter 2 is a region of subduction volcanism sur- rounding the Pacific Ocean. Of all nations, Indonesia is the volcano champion. It is completely surrounded by volcanoes, in all directions. Central Asia, where the Indian plate still pushes north, is not far behind—especially Kazakhstan, Pakistan, Afghanistan, and Iran. The American west coast from Chile to Alaska, part of the Ring of Fire, is very active. Alaska alone has about 60 volcanoes, the rest of the United States only about 40, and few of them are active (Canada has up to 100, but only about 18 active or potentially active ones, mostly along the Pacific coast). Southern Europe is another active area. As the African plate rudely muscles in on Europe, Turkey, Greece, and Italy are the most vulnerable.[1] Vesuvius is the only currently active volcano on the European mainland; Mount Etna in Sicily is the largest and highest volcano in the region.

Volcanic eruptions can be gigantic. Since historical times (counting from say 10,000 BC) about 50 eruptions have ejected more than 2.4 cubic miles of ash and rock. If you go back farther, the eruptions can get much greater: Toba, the Sumatran eruption that nearly extinguished our species 74,000 years ago, threw up maybe 120 cubic miles of debris; Yellowstone, some 700,000 years ago, about the same.

The Greatest Volcanic Eruptions in Recent History

Santorini, 1470 BC, 2.4-plus cubic miles of debris, destroying the Minoan civilization;

Taupo (New Zealand), 186 AD, 19-plus cubic miles, with pyroclastic flows traveling more than 62 miles;

Ilopango (El Salvador), 260 AD, 2.4-plus;

Rabaul (New Guinea), 536 AD, 2.4-plus;

Hekla (Iceland), 850 AD, 2.4-plus;

Baitoushan (China), 1010 AD, 36;

Laki (Iceland), 1783, 0.24 (but largest recorded lava flow);

Tambora (Indonesia), 1815, 36-plus;

Krakatoa (Indonesia), 1883, 4.8-plus;

Katmai (Alaska), 1912, 2.4-plus.[2]

Most volcanic eruptions, in point of fact, kill no one at all. It has even been argued that on balance volcanoes are beneficial. They serve as pressure release valves for all that molten rock, for one thing. Over the long haul, in geologic time, they have played a major role in modifying the planet on which we live, to make it more acceptable for humans. The United States Geological Service (USGS) estimates that more than 80 percent of the Earth's current surface, above and below sea level, is of volcanic origin. Better, the gas emissions from volcanic vents over hundreds of millions of years were instrumental in forming the earliest oceans and atmosphere, which supplied the preconditions for life. Volcanic eruptions have thrust up mountains, plateaus, and plains, which subsequent erosion and weathering have formed into fertile soils.

Volcanic soil is, at least after a time, immensely fertile. So much so that humans soon colonize the surrounding areas after an eruption, thereby increasing population density around the volcanic core—and of course increasing the hazard to themselves over time. As the USGS puts it, "Mankind must . . . learn not to crowd volcanoes . . . [but must] live in harmony with them and expect, and should plan for, periodic violent unleashings of their pent-up energy."

The agency has also listed other "benefits" of volcanoes, a list that includes geothermal power (energy generated from The Geysers geothermal field in northern California can meet the power needs of San Francisco); hydroelectrical power (water that flows down mountainsides, which seems to be a bit of a stretch as a "volcanic" benefit); health spas and hot springs (how else Baden Baden, Japan's Beppu, Canada's Banff Springs?); as a source for metals such as gold, silver, molybdenum, copper, zinc, lead, and mercury brought up from the depths; as construction materials (aggregate for road construction and cinder athletic tracks, cut blocks from hardened volcanic ash for buildings and walls, water-resistant concrete from volcanic ash); as a source of diamonds and opals; and a raft of miscellaneous items, some of them an even longer stretch as "benefits": as a source of drilling mud for wells; for kitty litter; for stone-washed jeans (pumice); as abrasives in toothpaste and kitchen cleansers and for cleaning dirty sinks. There is even a suggestion that the burial of human communities (Pompeii) was a contribution to studying history, which takes stretching the usefulness of volcanoes to extravagant new lengths. . . .

In a slightly more ominous way, volcanoes can help reduce global warming, though it would behoove us not to count on it. But a recent study in *Nature* by P.J. Glecker and others demonstrated pretty clearly that nineteenth-century ocean warming and sea-level rise were substantially reduced by the colossal eruption in 1883 of Krakatoa in the Sundra Strait in Indonesia—volcanically induced cooling of the ocean surface penetrated into deep layers, where this cooling persisted for decades after the event, much longer than expected, and was sufficient to offset a large fraction of the ocean warming caused by anthropogenic influences. The 1991 Pinatubo eruption, similar in size to Krakatoa, had a comparable effect but it was much shorter because of the much greater background warming in the twentieth century.[3]

Volcanoes differ from other mountains. "Normal" mountains are the product of shifting tectonic plates and the consequent crumpling of the crust. The Himalayas came into being as the Indian plate pushed its way northward; the Rockies from the collision of the American plate and the Pacific plate. Volcanoes, by contrast, make themselves up as they go, building up by the accumulation of their own eruptive debris, variously lava (which is what magma or molten rock is called when it reaches the air), tephra (airborne ash and dust), and "bombs" (crusted ash flows). The creation of a volcano can happen very fast, on a time scale alien to the ponderous movement of the plates. To take one of many examples, in 1943, near the village of Paricutin, Mexico, the ground suddenly opened on a cornfield belonging to the peasant farmers Dionisio Pulido and his wife, Paula. Paula, who had been watching the family's sheep, saw the whole thing start, when what looked like a tiny whirlwind sprang up in a field, and a small crevice, only a few inches wide, opened in the soil. Dionisio saw a small puff of green smoke. A few minutes later there was a rumbling, and a gray ash puffed up from the hole. Then the ground swelled up like a baking cake and fell back to the Earth. The family fled. By the following day, the cone was 33 feet high, and a hill of stones was oozing lava. A geologist, Ezequiel Ordóñes, was dispatched from Mexico City, and knew at once he was witnessing a very rare event, the birth of a new volcano. By June, the whole village had to be evacuated, by October the surrounding villages too, and by July only one street in the provincial capital, San Juan Parangaricutiro, was free of lava. A year later the volcano, now named Paricutin, after the village where it first started, was more than 1,400 feet high, and more than 96.5 square miles had been buried in ash. Then it went quiet. Ten years after it started, it was regarded as extinct. It had killed a few cows and wildlife, but no people.[4]

Most, but far from all, volcanoes are the classic conical shape of Japan's Mount Fuji. They are essentially vents that connect with reservoirs of molten rock deep below. The rock is forced upward, finds a weakness in the crust, and breaks through. It may just ooze to the surface, but if the vent is still plugged by solid rock, a violent explosion may occur. All magmas contain gases of various compositions, which can be liberated quietly or explosively, throwing out great chunks of rock and clouds of ash. Some of the leftover "rock froth" is called pumice, which can be light enough to float.

For convenience's sake, scientists have created a neat taxonomy of volcanoes, though nature is seldom quite so orderly. The main types are cinder cones, composites, shields, and lava domes. Paricutin was a classic cinder cone, with a neat bowl-shaped crater at the summit, built from a single volcanic vent and consisting largely, as you might expect from the name, of cinders that fell to Earth during the eruption. Dozens of cinder cones are found in every volcanic region on Earth. They seldom reach heights of more than 1,300 feet or so. Composite volcanoes, or stratovolcanoes, are also cone shaped and symmetrical, though more complex in their makeup, and are often much bigger, rising to 8,200 feet or more, composed variously of lava, ash, cinders, boulder fragments, and "bombs." Mount Fuji is the iconic composite. Composites grow as lava rises from the deep. Sometimes it flows through cracks in the cone's wall. It may then form ribs, which strengthen the cone. But those same ribs may block further flow, building up tremendous pressure. This is what happened when Mount St. Helens exploded in 1980. Instead of venting through the cone, the mountain effectively blew out sideways. Scientists had watched the mountain bulging but didn't really understand what was happening, resulting in a much greater loss (57 people killed and more than a billion dollars in damage) than should have been the case.

Extinct composites erode away slowly—Crater Lake in Oregon was once a high volcano named Mount Mazama, probably as high as Mount Rainier. The depressions left by eroded composites are called calderas. Ngorongoro is such a caldera; so is Yellowstone National Park, possibly the largest on Earth.

Shield volcanoes don't make mountains, only flattish domes that do resemble a warrior's shield, with a little imagination. They are usually not explosive but are built through thousands of lava pulses, each one laying a thin layer of basalt rock on the surface. Some of the world's largest volcanoes are shields—the Hawaiian islands are a linear chain of such volcanoes. The Siberian Traps and the Deccan Traps in India are startlingly large examples of volcanic plateaus rather than cones, but there are dozens of others, in Iceland, across northern Europe, and in western North America, including the Columbia River plateau, which erupted maybe 17 million years ago and covered 130,000 square kilometers. (The word "trap" is derived from the Germanic "treppe," or the Swedish "trapp," both meaning *stair*.)

Lava domes are made of lava too thick to flow very far. They grow from expansion from within, and because the sluggish lava typically hardens in the vent, they grow by shattering the cap and exploding violently outward and upward. Mount Pelée in Martinique, which exploded in 1902 to horrific effect, was a lava dome.

Eruptions are often described with labels that refer back to the archetype—Strombolian eruptions, Vulcanian, Hawaiian, Peléean, and, in a nod to Pliny himself, Plinian, which are often the strongest of all (Mount St. Helens was a Plinian eruption). Paricutin was a Vulcanian eruption; the Mayon eruption in the Philippines in 1968 was a Peléean eruption, named after that of Mount Pelée in the Antilles. Such eruptions are sometimes accompanied by pyroclastic flows, a technical term for deadly streams of "fluidized" hot ash, rock fragments, and scalding gases sometimes called *nuées ardentes,* or hot gases, deadly clouds of debris that can move at astonishing speeds (up to 310 miles an hour on occasion) and that have been known to travel up to 62 miles from their source, crossing large bodies of water as they do so. Sometimes these plumes can abruptly appear a great distance from the eruption itself. Mount Pelée produced just such a killer cloud. More than 27,000 people in the city of St. Pierre died, virtually the whole population of the town.

By a grim coincidence, Mount Pelée's eruption of May 8, 1902, was preceded by one day by a nearly identical disaster only 100 miles away, on the island of St. Vincent. There, the volcanic mountain La Soufrière exploded, devastating 45 square miles and killing 1,350 people. In both cases, the mechanism of death was the same: dense cloud of superheated gases traveling at highway speeds or greater, a pyroclastic outburst, a classic *nuées ardentes*.

Mount Pelée had begun to smoke and steam sometime in January, showing "fumarole activity," as the geologists put it, but nothing much else happened for a few months. Then, in April, minor explosions could be heard at the summit, and the city of St. Pierre, until then known to tourists as "the Paris of the West Indies" and home to no fewer than 16 rum distilleries, was shaken by a series of tremors. More than 50 people in the town were bitten by poisonous insects and venomous snakes that fled the mountain's slopes. Fiery red ants streamed off the slopes and through the town. Etang Sec crater lake nearby heated to boiling point, and on May 5 the rim gave way, sending a torrent of scalding water down the Blanche River, overrunning a distillery and killing 23 workers before cascading into the sea at 62 miles an hour, causing a 10-foot tsunami.

Politics made things worse. A committee of dignitaries, none of them with any volcanic experience or even scientific credentials, hiked to the summit and declared that there was nothing to fear, "nothing in the activity of Mt. Pelée that warrants a departure from St. Pierre . . . the safety of St. Pierre is completely assured." The only people on the island rich enough to leave were generally supporters of the governor, and since there was to be an election on May 11, the incumbent cajoled the local newspaper into downplaying the problem. When some residents tried to leave anyway, he sent troops to turn them back. Residents who lived around the volcano took what they believed would be refuge in St. Pierre, and the town's population ballooned from 20,000 to 27,000.

There was no further warning. As Alwyn Scarth reports, at 8:02 in the morning, a businessman in Fort de France was on the telephone to a friend in St. Pierre. He had just finished his sentence when "I heard a dreadful scream, then another much weaker groan like a stifled death rattle. Then silence."[5]

The volcano had erupted a few minutes before with a deafening roar. A massive black cloud of superheated gas, ash, and rock poured down a V-shaped notch in the mountain's south flank at more than 100 miles an hour, striking St. Pierre less than a minute later. Its impact toppled a 3-ton statue and crumbled meter-thick masonry walls into rubble, igniting huge bonfires of rum in the distilleries. Temperatures reached more than 1650 degrees Fahrenheit, hotter than the molten rock in the *nuées ardentes* cloud itself. A pyroclastic torrent swept through the town and into the harbor, destroying all but 1 of the 20 or so ships anchored offshore. Among them was the American sailing cruiser *Roraima*, which had arrived for a visit only hours earlier.

Where the *nuées ardentes* swept through, only 3 people survived, a shoemaker, a criminal in his subterranean jail cell, and a little girl who took refuge in a small boat in a sea cave. (The felon, Louis-Auguste Cyparis, survived for 4 days before he was rescued. He later was pardoned and joined the Barnum & Bailey Circus, where he toured the world billed as the Lone Survivor of St. Pierre.) The rest died in the force of the blast or from inhaling superheated fumes and ash. Elsewhere in the region, a few survived, lucky by a few miraculous feet.

That October, a lava dome rose from the crater of the ruined mountain. It grew for a year into an obelisk 492 feet thick at the base and nearly a thousand feet tall, accreting more rock than contained in the Great Pyramid at Giza, growing at about 50 feet a day before it collapsed into a pile of rubble in March 2003. There it lies, quiescent and safe, it is thought. For now.

Some active parts of the volcanic world are more placid than this. Hawaii, which is self-evidently volcanic—the existence of all the islands clearly the product of volcanic eruptions—has always been a good test of the theory of plate tectonics and even better evidence for the existence of thin-crust hot spots. This is as good as it gets for a volcanologist: you can study the volcanic process while living directly, or as directly as you care to, on top of the volcanoes in question.

The indigenous Hawaiians knew, by tradition and legend, how Hawaii was formed; though they lacked the vocabulary, they knew how to recognize the difference between the older islands, now heavily eroded, and the more recently created and un-eroded ones to the southeast. In Hawaiian myth Pelé, the goddess of fire, was forced from island to island as she was ousted by a succession of hostile deities, her increasingly desperate journey punctuated by massive fiery explosions, until she finally came to rest on the relative sanctuary of Big Island. Her island-hopping journey corresponds neatly with the modern theory of how and when the island chain was created.

The theory is, essentially, this: on the sea floor under the Hawaiian chain is a planetary hot spot, which is just a shorthand way of identifying a weakness in the lower crust and upper mantle that lets plumes of magma thrust toward the surface. This hot spot, like all such, is independent of the ever-shifting continental plates, being deeper than they are. So when an eruption comes, as it inevitably does in long geologic time, the resulting magma plume builds toward the surface of the sea. Eventually, in maybe fifty thousand, or a hundred thousand, or a quarter of a million years, it pops above the water, and after an epoch or two birds will find it, and weeds, and coral, and the slow but inevitable colonization will begin.

The islands age, as the Pelé story suggests, in a more or less straight line from southeast to northwest. There are only two plausible explanations: either the crust is moving northwest, or somehow the internal convection chamber is itself finding a linear succession of spots through which to erode. Most scientists now accept that the crust itself is moving across the hot spot, so that the islands erupted and accreted in sequence. Hawaii is in the center, more or less, of the Pacific plate, which is drifting northward at about ten centimeters a year. All the islands were independently produced. They are not connected to each other, deep down.

The process is ongoing. Big Island is still growing. Only a few miles off the southeast coast is an underwater seamount called Lo'ihi, and someday soon, in a piffling fifty thousand years or so, it too will peek above the waves, and there'll be room for a few more dozen surfer subdivisions. For the most part, the process has proceeded peaceably enough. The basalt lava of which Hawaii is made is highly viscous, charged with gas, and flows easily rather than exploding. Its pace is often leisurely. A trotting donkey can easily outrun it.

Europe has had a more turbulent volcanic past, as Pliny's Vesuvius would suggest.

A cluster of volcanoes, like Vesuvius itself and Etna, string a line along southern Italy, Greece, and into Turkey, a consequence of African colonization—the African plate is pushing steadily into the Aegean plate at the stately pace of perhaps 2 inches a year, thrusting up mountains, shaking the Earth, and venting great lakes of magma in volcanoes. It has been happening at least since early human history, and perhaps before that. The whole Bay of Naples is built of Campanian Ignimbrite, "whose origin was in a titanic explosion some 35,000 years ago that would have dwarfed Vesuvius. It blew out a great hole in the coastline of Italy, at the Tyrrhenian Sea, and a vast cloud of incandescent material was carried along. There were probably Paleolithic witnesses to the destruction."[6]

Geologists like Fortey think the Campi Phlegraei, near Naples, is potentially more dangerous than Vesuvius. A magma chamber 1.25 miles below the surface is rising—"and some volcanologists are deeply worried that it, rather than Vesuvius, where the magma chamber is 3 miles deep, may be about to explode." The Monte Nuovo eruptions of 1538 might have been a preamble of what is to come. Monte Nuovo, in the middle of Pozzuoli, the leading trading center of the region, erupted after the surrounding land had been swelling ominously for decades. It was a small eruption and showered Pozzuoli with a muddy brown ash, but that's no guarantee it won't be larger next time. From 1969 to 1984 the cone of the old volcano—in the middle of Pozzuoli harbor—has bulged by 8.2 feet; earthquakes in 1982 were worrying enough to prompt a partial evacuation of Pozzuoli, a thing hazardous enough in itself, given the crowding and the state of Neapolitan traffic.[7]

Mount Etna, which looms over Messina and eastern Sicily, is the largest active volcano in Europe, some 11,000 feet high. It is mostly benign, but every now and then its choler seems inexplicably to rise. Its most famous bout of ill temper was during the Ides of March in 44 BC, just as Brutus and his cronies were putting paid to the great dictator, Julius Caesar, on the steps of the Senate. The sky went dark and remained so for more than a year, a gloomy portent and a sign the gods thought the killers ill-favored—or so believed the mob that bayed after Brutus in the murder's aftermath. In that eruption, debris clouds from Etna fell on Italy, Greece, and Egypt, causing famines in all three countries; the fleeing assassin, Mark Antony, was forced to eat "roots, bark, and animals never before tasted by men."

Virgil, rising to the full heights of oratorical purple in Book III of *The Aeneid*, described the mountain thundering with terrifying crashes.

> The port capacious, and secure from wind,
> Is to the foot of thund'ring Aetna join'd.
> By turns a pitchy cloud she rolls on high;
> By turns hot embers from her entrails fly,
> And flakes of mounting flames, that lick the sky.
> Oft from her bowels massy rocks are thrown,
> And, shiver'd by the force, come piecemeal down.
> Oft liquid lakes of burning sulphur flow,
> Fed from the fiery springs that boil below.
> Enceladus, they say, transfix'd by Jove,
> With blasted limbs came tumbling from above;
> And, where he fell, th' avenging father drew
> This flaming hill, and on his body threw.
> As often as he turns his weary sides,
> He shakes the solid isle, and smoke the heavens hides.[8]

In 1669 Etna did it again. This was a slow-moving eruption, lasting for months and oozing lava out at the stately pace of about a hundred meters an hour, ample time to move out of the way, and even to move possessions. The lava flowed through the town gates of Catania, but most of it was turned by the town walls and flowed along them to the sea.

The flow halted for a week, by happy chance Easter week; the church seized the opportunity to perform the necessary exorcisms, to not much avail, for the flow started up again. By the end, in July, about a half mile of lava 80 feet thick had oozed from the mountain and flowed about 22 miles down the coast. No one was hurt or killed, and most of the townsfolk watched its progress from the town walls. It became something of a game.

You don't have to actually go to Santorini, on the other side of the Italian Boot, to see the place that was the royal center of the probably mythical or might-have-been sea-going empire of Atlantis, which once rivaled Athens for mastery of the ancient world. You can see it on all the official Greek tourist posters, for one thing, a glorious shot of a picturesque whitewashed village clinging to a cliff three hundred meters or more above the azure sea—the number of tourist snaps of that village must surely rival the number of years in geological time or the hamburgers sold by McDonald's. You can do better than that now: you can watch in real time, on the web, the caldera that was once a mountain that once might have exploded to totally destroy that might-have-been empire . . . a webcam now watches the crater, refreshed every sixty seconds, just in case.[9] But if you only watch the webcam instead of going to Santorini, you would be doing yourself a great disservice, for if you caught the ferry from Rhodes or Mykonos and went there, you could wander down to the beach in the little village of Fira (Thira, or Thera) and watch the little five-vessel fishing fleet come in at dawn and then again at dusk, and then an hour later you could wander into one of the tavernas, have a glass of whatever pleases you and eat a plate of silky sardines brushed with olive oil from the mountains above and grilled over charcoal, could sit at the taverna table, and stare across the small bay at the misshapen rock called Nea Kameni, or New Burned. Which once exploded and began the whole thing.

On a quiet evening, when the sardines are gone and the fishermen abed, you can listen to the volcano hiss. It doesn't sound evil. It's just a hiss. But long ago it exploded, and the Minoan civilization, once a glory of the ancient world, disappeared into oblivion.

That was, they say, 3,634 years ago, counting backwards from 2008.

Santorini is five islands now, where once it was one. The 29 square miles of these five small islands is less than half of what the island was before. The caldera, or crater, is 20 miles across. The present village of Fira, or Thera, clings to the edge of the cliff the explosion created, straight down into the sea, 984 feet below; until just a short while ago, the only way to get from the top to the shore was via a precipitous and vertiginous path carved into the cliff face, used daily by insouciant donkeys. Halfway down you can get a good view of the remnant volcano, Nea Kameni. It doesn't look dangerous, despite its *sotto voce* hissing. You can wander around on it, only occasionally getting a whiff of sulfur, a reminder that it is not, quite, finished with its work.

In the old days, Thera would have been a modern Bronze Age city. Archeological evidence, excavated through more than 66 feet of volcanic ash, shows a town with running water in cisterns and pipes, public and private baths, multi-storied buildings and homes, and no evidence whatever of military materiel. No human skeletons have been uncovered in the ruins. Perhaps the inhabitants had sufficient warning and fled the island before it blew. A few might have gone to Crete and perished there instead. A few might have gone to the Italian mainland, North Africa, or even Palestine—Judaic scripture says the Philistines came from Caphtor, or Crete.

The explosion would have had many consequences. First were the tsunamis. The Cretan tsunamis would have come in at speeds approaching 185 miles an hour, at heights that would have been between 100 and 300 feet; the Minoans were a seafaring nation and clustered around the shore—all their villages, and all their fleets, would have been engulfed and destroyed, along with all the skilled workers. The second catastrophe would have been the ash fall. Most of Crete and large parts of Turkey were covered with ash from the eruption. It takes very little ash to put a stop to plant growth. Centimeters will prevent photosynthesis. A few inches will kill ground vegetation. Only a little more will poison the soil for years. A few feet will make it impossible to grow food for decades. In this case, there were feet on the ground.[10]

For all the U.S. Geological Service's valiant attempt to domesticate volcanoes and to persuade us that they are, on balance, lovely things and quite benign, half a dozen times in human history volcanoes have critically affected global temperatures, and therefore global climate, and therefore human life.

The four best documented, because they happened in modern times, were the eruptions at Laki, Iceland, in 1783; Tambora,

Indonesia, in 1815; Krakatoa in 1883; and Pinatubo, Philippines, in 1991.

The Laki eruption started on Whit Sunday, the second Sunday in June, in 1783, and the reason we know a good deal about it is due to the notes made by Pastor Jon Steingrimsson and a few of his parishioners, who were the first to see a plume of black smoke rising over the horizon, followed by loud groaning and rumbling sounds, and a cloud of ash.

Steingrimsson was too sensible a man to believe in evil portents, but the results were evil enough: the Skafta River abruptly stopped running, and instead a wave of steaming lava came rolling down the riverbed, moving at about six kilometers an hour. The rain that followed was highly acid, burning the crops and choking sheep.

The eruption lasted more than 8 months, ejecting a 0.24 cubic miles of ash into the atmosphere and more than 3.35 cubic miles of basalt lava. A blue haze of toxic gases, fluorine, and sulfur dioxide hissed from the crevices surrounding the mountain and entered the jet stream; within days 80 to 100 megatons of gas spread across Europe and as far as Syria. Ben Franklin, who was in England at the time, thought the cause might be a volcano but also suggested another possible mechanism: maybe the Earth had passed through a cloud of cosmic smoke?

Obviously, the effects were worst on Iceland itself, where most of the island's livestock died from eating grass contaminated with fluorine and most of the next year's crop failed because of the acid rain, and a quarter of the population perished in the resulting famine. The event also caused the Haze Famine in Ireland, and across Europe an unusually hot July followed. The next winter, however, was 39.7 degrees Fahrenheit below the 250-year norm across all of the northern hemisphere. The blue haze persisted for three more years, causing three more cold winters, before the natural pollution diminished and the climate returned to something approaching normal.

The second of the catastrophic volcanoes was the abrupt explosion of Mount Tambora, which dominated the northern end of

Sumbawa Island, one of a chain of volcanic islands stretching from New Guinea to Sumatra. The whole chain is geologically active and under intense pressure as the southeastward-moving China plate passes over the northeastward-moving Indian plate; it was this same tectonic drift that set off the earthquake that caused the Christmas Eve tsunami of 2004. When it erupted, it was with the most violent explosion in recorded history, or at least the last 10,000 years. The noise was heard more than 900 miles away, and about 36 cubic miles of rock was hurled into the sky, reducing the height of the mountain itself by 4,200 feet. Because it took place above-ground, it did not generate a tsunami, but the material that went up stayed up for a year.

No official census of the disaster was ever taken and the "official" death toll of 12,000 found in most of the reference books is no more than a guess. Recently, however, the estimates have been raised dramatically, as a team of volcanologists uncovered evidence of the lost kingdom of Tambora, which had been known throughout the Indies for its honey and wood products, but which vanished entirely in the eruption. The team, led by Haraldur Sigurdsson of the University of Rhode Island, suggests that more than 90,000 perished in the initial blast.[11]

Its effects were catastrophic far beyond the Indonesian Islands. The event generated an extended period of global cooling; as far away as New England the climate cooled dramatically. The following year, 1816, was called "the year without a summer"; snow fell in June, then again in July, and even August. Most crops failed, as they did in Europe, and possibly another 90,000 or so died as a result. No one associated the plunging temperatures with Tambora, then. As already reported, public indignation, at least in Massachusetts, was briefly aimed at the omnipresent Ben Franklin, whose experiments with attracting lightning through high-flying wires attached to kites were popularly thought to have punched holes in the sky to let the frozen wastes of space into the lower atmosphere. The annual temperatures recorded at Yale for 1816 showed temperatures 44.6 degrees Fahrenheit below the norm.

The explosion of the Indonesian mountain of Krakatoa in 1883 may be the most famous in history, after the one that buried Pompeii. Partly this is because it was so well documented, but partly also because its effects were felt so far away. And partly, no doubt, because at least in English its very name sounds violent.

The eruption wasn't as large as Tambora (maybe 4.8-plus cubic miles of ejecta, compared to Tambora's more than 36) and didn't kill as many people, but it was the second largest in history, did cause oddly cool weather globally for months afterwards, and did lead to brilliant sunsets; part of its legacy is a collection of marvelous paintings of the Thames River by artists like J.M.W. Turner, whose paintings gave birth to the Impressionist movement.

The mountain came awake in May. On the twentieth there were mild enough eruptions from a summit named Perboewatan, which was pretty much destroyed; several new vents appeared near Danan, one of the cones of the complex mountain. Mild, but not insignificant: early that morning the German warship *Elizabeth* reported seeing a seven-mile-high cloud of ash and dust rising above the island. The earliest debris ejected was basalt, which seemed to indicate that the magma chamber far beneath was being recharged; a month later, lighter pumice was found floating throughout the Sunda Straits.

For the next two months passing vessels were entertained with similar black and gray clouds of ash and their associated rumblings. No one took these too seriously. Indeed, explosion-watching festivals were held in a number of places on the safely distant (they thought) Javan and Sumatran coasts, and cruise ships were hired to take tourists to see the fireworks.

At 12:53 P.M. on Sunday, August 26, Krakatoa showed them why this was a very bad idea. The opening salvo was a massive and ear-splitting roar that pulsed a black cloud almost 22 miles into the air; within hours the coastal communities of western Sumatra, Java, and nearby islands were swept by a series of massive tsunamis and clouds of hot pyroclastic ash.

Even this wasn't the worst. Early the next morning there was a series of four enormous eruptions, climaxing in an explosion that simply tore Krakatoa apart; 89.5 square miles of the island collapsed into the now-empty caldera. Ernest Zebrowski has written, "The noise was heard 4,600 kilometers [2,850 miles] away, from Sri Lanka to Australia.

Residents of Darwin n Australia were woken by the blast; at Diego Garcia, 3,647 kilometers [2,266 miles] away, the sound was mistaken for cannon fire. Rumbles were heard over one thirteenth of the Earth's surface. . . . So violent were the explosions that they actually set the Earth's entire atmosphere ringing like a bell."[12] They caused an atmospheric pressure wave that was recorded in many locations remote from the volcano—France, England, Alaska, Hawaii. Although the direct tsunami waves couldn't have reached the American west coast (they would have been "shadowed" by continents and islands), an atmospheric gravity wave set off perturbations that indirectly transferred energy to the ocean—causing a 6-inch surge in Sausalito, California.

Within minutes, hundreds of coastal villages were destroyed and some 37,000 people perished in 100-foot tsunamis that swept in at 250 miles an hour. Auckland, New Zealand, 4,826 miles away, recorded a 6.5-foot wave a day later.

Another 4,500 people died in the falling debris called tephra, and pyroclastic flows that crossed the sea and swept up low-lying shores, sometimes traversing 25 miles of open water, leaving corpses with scorched lungs and burned flesh; it was still hot enough to incinerate whole villages. It reached even farther: the steamship *Louden* was 40 miles north-northeast of Krakatoa when it was struck by strong winds and falling tephra, and the *W.H. Besse*, even farther at about 50 miles, was hit by hurricane-force winds, heavy tephra, and the strong smell of sulfur.

As with Laki and Tambora, the effects were felt far beyond the volcano's immediate neighborhood. Dust and ash and sulfur dioxide gas were driven high into the stratosphere, spreading across the entire equatorial belt in only two weeks. The sulfur dioxide combined with water to make sulfuric acid aerosols that reflected enough sunlight to drop global temperatures by up to 39 degrees Fahrenheit for months afterwards.

It may also be that Krakatoa is not quite done. In 1927 a small flotilla of Javanese fishermen was crossing the now-collapsed caldera when the men noticed steam rising from the sea, accompanied by small bits of debris. A year after that the uppermost rim of a new cone appeared above the water, and a year after that it had grown into a small island, which the locals came to call Anak Krakatoa, Child of Krakatoa. It hasn't done anything spectacular since then, but nor has it stopped. Most years there are a few modest eruptions, and pulses of basalt lava emerge. They don't pose any danger, or apparently don't pose any danger, to surrounding villages. Local residents are keeping a wary eye out anyway, lest the child grow into a tantrum-tossing adult.

June 1991 was not a good month for the region around Pinatubo, in the Philippines; in the middle of the month, Typhoon Yunya descended on the area with howling winds in excess of 125 miles an hour, just as Mount Pinatubo itself, which until the previous year no one had even known was a volcano, blew up.

This was the second-largest eruption of the twentieth century and produced the largest sulfur dioxide cloud. As a consequence, mean global temperatures went down by 2 degrees for over a year.

Despite the confluence of calamities, no one was killed, though the surrounding area contained more than half a million people. This was in part because it was only a few miles from Clark Air Force Base, a U.S. facility, and those alert gentlemen spread the word to the neighboring towns and villages, and in part because the mountain itself erupted in an orderly fashion—it had been set off by an earthquake late the previous year and erupted in a sequence of ever-increasing blasts, from small hissings to the final paroxysm. Nevertheless, several towns had to be evacuated, and thousands of small buildings collapsed from the weight of the ash that was subsequently moistened by torrential rains dropped by the typhoon.

It was the eruption of Pinatubo, as we shall see, that gave rise to one of the more fanciful, not to say controversial, suggested "remedies" for global warming.

Yellowstone National Park is, potentially, the greatest super-cataclysm of them all.

This is no great secret. The park's own literature will cheerfully point all this out to those millions who go there to see the remarkable landscapes, including the park's geysers, fumaroles, and hot springs, with their associated improbably colored rocks. Many of the millions of people who have watched Old Faithful geyser spurting upward every sixty-five minutes or so go home with a pamphlet explaining in great detail what happened in the past, what is happening now (as far as they can tell), and what might, just possibly, happen in the future. And what is this thing that might happen? In the geological short term, a mere eyeblink of time, it will almost certainly explode, devastating most of North America and changing the climate of the Earth for centuries to come. No one seems to panic. It might happen tomorrow, but on the other hand it might not happen for, well, millennia.

The journal *Nature* has described Yellowstone this way:

> [It] is the youngest of a series of large calderas that formed after sequential cataclysmic eruptions that began about 10 million years ago in eastern Oregon and northern Nevada. The Yellowstone caldera was largely [created] by rhyolite lava flows during eruptions that took place about 150,000 to 70,000 years ago. Since the last eruption Yellowstone has remained restless, with a high seismicity, continuing uplift/subsidence episodes with movements of some 70 centimeters to several meters since the Pleistocene epoch, and intense hydrothermal activity. . . . The current movement probably results from variations in the movement of molten basalt into and out of Yellowstone.[13]

It is the "high seismic activity" that gives visitors to Yellowstone their *frisson*.

Yellowstone is what is called a "resurgent caldera"; such calderas are typically the largest volcanoes on Earth, associated with massive eruptions on a scale not (yet) seen in modern times. Yellowstone is not the only one, though it is probably the biggest. Mount Toba in Sumatra is the youngest—it erupted some seventy-four thousand years ago, in an explosion five hundred times larger than Mount St. Helens, and much greater than any volcano in historical times—it was Toba, as already reported, that nearly did in the human race. Two other resurgent calderas exist in North America—the Valles caldera in New Mexico and the Long Valley caldera in California. Mammoth

Mountain, that ever-trendy skiing community, is at the western edge of the Long Valley caldera, and in recent years anxiety-inducing activity has been seen there—fumarole gas emissions began in the mid-1980s, accompanied by shallow-level earthquakes, which indicate the magma is rising. Carbon dioxide emissions seeping through cracks have killed trees in several areas around Mammoth.

The existence of the enormous Yellowstone caldera was suspected all along but only recently identified. For years, geologists, knowing the region was volcanic, had been puzzled that there was no obvious caldera in the park area; it wasn't until NASA satellites mapped the area that they could see why they couldn't spot it—the whole park is a single caldera nearly 35,000 square miles in extent, underlain by a magma chamber 7.5 miles thick. If it ever blows (and that's "if"—don't reach for your hard hats yet), it would probably do what the last really big one did 640,000 years ago, which is to darken the sky for generations, destroy most wildlife, at least in western North America, and lay down enough ash to cover the entire western United States and Canada in a layer more than a 3 feet thick. (It would be enough to cover all of New York State in 230 feet of ash.)

The park is still very active. In 2002 the geysers started spurting a little faster than before (neatly coinciding with an earthquake in Alaska, which showed how much more interconnected the underlying magma was than previously suspected). In 2007, using radar interferometry, researchers identified an unusual feature, the bulging upward of a large area about 20 to 25 miles centered under the north rim of the caldera. Even more unusually, during the uplift episode the caldera floor had subsided. "This type of deformation has not been seen before and has profound implications for the workings of large calderas," the same *Nature* report said.[14] Clearly the massive reservoir of magma is moving. One of the precursors of a volcanic explosion is increased seismic activity—and Yellowstone is currently experiencing somewhere around a thousand tiny quakes every year.

Still, it might not blow at all. Just because something has happened before, doesn't mean it will happen again. Just because Yellowstone seems to have "blown" every 600,000 years or so, and just because the last one was some 640,000 years ago, doesn't mean that it is somehow "overdue" by 40,000 years. As with asteroid impacts, volcanic eruptions are a stochastic process, non-linear events with no inbuilt timetable.

The increased geyser activity might even mean that the super-volcano is venting its more volatile gases, which would make an explosion less likely.

In truth, no one knows for sure. And if they did, there is nothing that could be done about it anyway.

TEN

Poisonous Emissions and Noxious Gases

But mother, please tell me, what can those things be
That crawl up so stealthily out of the Sea?
—English nursery rhyme

L ate on an August night in 1986 in the Cameroonian highlands a
deadly thing came out of one of the lakes. Whatever it was, it made
no noise. It had no shape. You couldn't see it or smell it. You couldn't
feel it coming. It rolled out of the lake and up over the escarpment,
finding the winding valleys where the villages were. It drifted down the
valleys, and came to the villages of Nyos, Kam, Cha, Subum, the last as
far as 14 miles from the water . . . and there it killed, quietly and without fuss. No one fought. No one screamed. No one even knew it was
there. Afterwards, a few Fulani herdsmen said they had heard strange
gurgling coming from the lake in the afternoon and saw the water was
fuming, as though it were on fire. A few people said later they'd heard
banging sounds, as though the Fulani and the Bamun were firing at
each other again, an old tribal feud. Most of the villagers just went to
sleep, but they didn't waken. Eighteen hundred people died that night,
six hundred in the village of Nyos alone.

A few survived, though they couldn't explain why their neighbors died and they did not. Their stories were recounted in a pamphlet I picked up a decade later in nearby Foumban. Some of them remembered suddenly not being able to breathe, and falling over, waking up eight hours later with a headache and aching limbs. A few recalled picking something sticky off their arms and legs; a few reported burns where there was no fire. Some had other gashes, small wounds, but had no idea how they got there. Rain was falling, but it burned the skin. Most just felt exhausted and fell back to sleep. They awoke in the afternoon to find all their neighbors dead. The cattle were dead too, and the goats. So were the chickens. For some days afterwards, no birds sang.

The turquoise water of the lake had turned bright orange and was choked with clumps of rotted vegetation and duller red patches; wisps of gas that might have been steam coiled upward from the southern part of the lake, where the rocks had been scrubbed bare of bushes.

The Cameroonian countryside changes rapidly as you recede from the rain forests along the coast. It becomes hilly, with large plantations of bananas and pawpaws. As the road ascends it begins to cool off and become much less humid, rather like California only with more rain. That's where the coffee is grown. South of Bafoussam every house has a fruit stall piled high with mangoes, pineapples, guavas, avocados, sugar cane, bananas, and roots of all kinds. At Bafoussam it begins to flatten out into plains. Cattle start to appear for the first time. By Foumban the countryside is still very fertile, but dry, and begins to resemble the Africa I know best, the Africa of the savanna. Every little house has a banana grove, and many a coffee tree or two.

From Foumban we turned west on a vertiginous road that skirted the Massif Mbam, more than 6,500 feet high, part of the volcanic chain that started in the south, at Mount Cameroon. Then to the picturesque town of Bangourain before joining the main road to Kumbo. This is where the lakes are. The landscape is stunningly beautiful, the ridges and valleys verdant with brilliant green vegetation punctuated by cattle the color of burnt sienna. The traditional Bamiléké architecture is distinctive, with sharply conical roofs in thatch; the new bourgeoisie are all building themselves villas in the high hills, roofed with aluminum, and the effect is unexpectedly alluring, green valleys lit up by flashes of silver. The lake country is of mixed ethnic heritage, partly Bamiléké and partly Bamum. Foumban, an hour or so north of Bafoussam, is the seat of the Bamum dynasty, one of the few kingdoms left in Old Africa from pre-colonial times. Until recently the kingdom was governed as it always had been, by the sultan and his council, who made wars and alliances and kept their little kingdom going long after the great African empires had collapsed in the face of European attrition.

Bamum is a peaceful part of a peaceful country. It seemed hardly fair that something should come up out of the water like a demon and strike the people dead. It was known, of course, in the more remote villages, that the lakes and streams were inhabited by a spirit woman of power. But why was she angry? Absent the notion of sin, a Christian import and not well regarded locally, there seemed no reason at all.

In the cities, and in the capital, they paid no attention to the villagers' stories. They just brought in the scientists.

It wasn't long before a flock of them descended on Lake Nyos—here was a new peril, unexplained. It needed to be understood lest other lakes, elsewhere, suddenly erupted. A team from France was first on the scene, in collaboration with scientists from Yaoundé, the national capital, closely followed by Americans, Japanese, and British.

It was known—or at least plausibly theorized—that the deadly cloud had been strong concentrations of carbon dioxide gas. Lake Nyos is in the Oku Volcanic Field, at the northern boundary of a zone of crustal weakness and volcanism. Lake Nyos itself occupies a crater that formed from a hydrovolcanic eruption a mere 400 years ago, well within the lifetime of the Bamum dynasty. It's not the only such lake: about 30 similar lakes are found nearby, all of them rich in CO_2—the lakes do, after all, lie over a volcanic source. Most of the gas doesn't dissipate but dissolves into the deepest levels of the water. At 650 feet, the sheer weight of the upper lake levels exerts a confining pressure, just like the screw cap on a gigantic bottle of Perrier. At those pressures, water can hold 15 times its own volume in CO_2. The CO_2-rich water is denser than ordinary water and tends to remain on the bottom; unlike salt, which is very stable in solution, carbon dioxide's solubility is limited by temperature. But this deeply buried soda water is inherently unstable.

There's no doubt the CO_2 is of volcanic origin. It seeps through the sediments at the bottom of the water column from a cluster of hot springs. And there's no question that a great deal of gas had escaped Lake Nyos during the "event"—the surface of the lake dropped more than three feet. That it killed so many people was an unfortunate consequence of the lake's geography. Since CO_2 is denser than air, it flows over the ground surface, asphyxiating people in its path; Lake Nyos is at an altitude of 3,280 feet and is drained by several long and fairly deep valleys, each of which shelters several small villages—and each of which, in turn, became a death trap instead of a shelter.

These lakes in Cameroon and in Rwanda are not the only parts of the world where CO_2 seeps to the surface in hazardous ways, which is one reason scientists paid them so much attention. Lake Averno, just up the coast from Naples near Pozzuoli, used after all to be considered one of the entrances to the underworld and emitted similarly seeping noxious gases.

If you stand on the Nyos escarpment now you can trace the path the gas would have taken. Of course, the water has returned to its normal deep blue, and the lake looks utterly beautiful. It is less than a mile wide and twice as long; the cliffs on the far shore are garish red, and brilliant green vegetation spills down every crevice to the water's edge. No one lives on the shores, not any more. The lake is always full—it rains constantly, more than 6 feet a year, and the surplus flows over a spillway cut into the northern rim of the crater and down a narrow valley, one branch of which leads to the village of Nyos.

The real question is—what set it off this time? What made it suddenly bubble to the surface? The CO_2 could have burst through the lake as the result of a sudden eruption deep below the lake. Or it could simply have been destabilized by the overturning of the bottom waters. This could have been done by a landslide, by an earthquake, or just by abnormally heavy rains and strong winds, which would have pushed the new water to the northern bank where it "slid" down the crater wall and disturbed the bottom water.

Whatever the cause, everyone agreed that both Monoun and Nyos still contained massive amounts of CO_2 (3.5 million cubic feet and 10.6 billion cubic feet respectively) and that they were constantly being recharged through hydrothermal springs, and that similar events could easily happen in the future. The "fix" was to degas the lakes through a kind of controlled limnic eruption. The idea was inspired by the industrial process called gas lift, used at many natural gas wells, and also by an existing degassing unit at Lake Kivu in Rwanda, another gas-bearing lake (where the intention is to use the resulting methane gas as a new energy source). A polyethylene pipe is installed between the lake bottom and the surface; a small pump raises the water in the pipe up to a level where it becomes saturated with gas, thus lightening the water column, which then rushes to the surface. Once the system is primed, the pump is no longer needed and the process becomes self-sustaining—the isothermal expansion of gas bubbles drives the flow of the gas–liquid mixture as long as dissolved gas is available.

Just such a column was set up at Nyos in January 2001, funded by the U.S. Office of Foreign Disaster Assistance (OFDA), the French Embassy in Yaoundé, and the Cameroonian government. It is remotely monitored and can be controlled from the capital, the data being transmitted via satellite. On January 31, a stable jet, 165 feet high, soared upward: the degassing of the lake had begun.

It is an extraordinary sight, this gas. If you stand on the shore it looks just like a fountain, powered by a powerful pump, no different in appearance from the dozens of fountains you can see in that great temple of wretched excess, Las Vegas. (Or, if you don't want to go all that way, you can watch the Lake Nyos fountain, live, on the web. One of the French scientists involved in the degassing project, Michel Halbwachs of the Université de Savoie, has set up a webcam that is refreshed twice every day; the accompanying data chart shows that the height of the column has been hovering around the 300-foot mark for the past few years. Halbwachs suggests that the degassing of the lake, at the current rate of 53,000 cubic feet a day, would be finished by the end of the decade.)[1]

L ake Nyos isn't the only part of the world to seep noxious gases, nor is CO_2 the only such gas so seeped.

The Pearl Hotel in Sochi, on the Black Sea coast, is a huge complex with its own beach, lots of chaises for lounging, umbrellas, bars, cafés, piles of watermelons in the shade of the jacarandas. I spent some time in Sochi in the last days of the Soviet Union, but it didn't feel very Soviet, being more suited to the children of the southern Mediterranean than to the stolid comrades of Novorossiysk. It's a subtropical resort, attractively sited, with lush green hills steep to the sea, snow-capped peaks not far away, health spas, holiday camps, resort hotels, clean stone beaches.

Twenty years later, terrible things had been done to the Black Sea. Nearby agricultural lands, in half a dozen neighboring countries, had been subjected to massive doses of fertilizer, much of which had run off into the sea, causing eutrophication; the infusions of nitrogen and phosphorus had changed the sea's ecology, caused massive die-offs and ghastly smells, and seriously damaged the fisheries and the tourism industry. The Black Sea was not quite (as one newspaper report put it) "a deadly soup of toxic wastes" but nor was it very pleasant, and as a result, the six offending Black Sea countries signed the Bucharest Convention in 1992, with the intention of cleaning up the sea and restoring it to health.

One of the things the commissioners discovered, during their inventory of the Black Sea's human-caused horrors, was something unexpected: a massive stratum of a perfectly natural substance called methane hydrate. Which explained something I had been told the first time I was in Sochi: A young couple at the Pearl's beach had told me that the sea sometimes caught fire. Indeed, they said, fishermen had often reported lightning causing fires on the water around their boats. At the time, I was inclined to dismiss the stories as just yarns (the same couple told me they believed the exotic dancers at the nearby cabarets were all KGB agents, which seemed an improbable use of that formidable agency's resources), though this was just after the Cayuga River near Cleveland, Ohio, had caught fire, so thick was it with industrial pollutants and solvents, so anything was possible. Now it seems likely: methane had spontaneously bubbled up from below the ocean floor and caught fire in the lightning or by agency of a carelessly tossed cigarette.

Methane, and especially methane hydrate, is one more dire peril facing humankind. Or it represents a solution to the End of Oil—it depends which authority you choose to believe. Possibly both extremes are true.

The origin of the methane is no mystery. It is biogenic, caused partly by the decay of plants and animals over millions of years, and partly by primitive bacteria-like creatures called Archaea, which seem to thrive in extreme environments—at undersea volcanic vents, where the temperatures are 212 degrees Fahrenheit or higher, in the geysers of Yellowstone, at extreme pressures, or in areas of extreme saltiness—even in the human mouth, as toxic an environment as any bug could wish. It's possible, in fact, that these Archaea, one of the three great orders of living things, have been around longer than any other extant life forms, nearly four billion years, dating to an epoch in which there was no free oxygen. Not surprisingly, they are sometimes called extremophiles. Some of these odd creatures eat methane, but many more excrete it—swamp gas is produced by Archaea. They do so in the human gut too. ("This is one of the gases that allows you to do your party trick, bent over, with a lighter in a darkened room."[2]) There are a lot of them, too—by one estimate, they represent almost a third of the biomass on the planet, perhaps more.

A gas hydrate, or methane hydrate, looks and feels just like water ice. It is made up of single molecules of methane trapped in a "cage" of six water molecules, and forms everywhere temperatures are low and pressures great. Hydrates can be found in really deep water, even below 10,000 feet, if enough sediment had been washed down there to decay, but most are on the continental shelves, between 1,650 and 3,300 feet. They are pretty common in nature, it turns out, in Arctic regions and in marine sediments. So common, in fact, that the Inuit on Ellesmere Island sometimes bash holes into the ground and set the holes alight—the escaping methane burns to keep them warm. Generally, wherever you have methane and water together at low temperatures, a hydrate will form.

A vast amount of the stuff exists. Hydrates are very efficient "containers" for methane; a liter of hydrate will contain 42 gallons of methane gas under pressure, all of which will be released if the hydrate melts. The worldwide amounts of methane and burnable carbon stored in hydrates is, by conservative estimates, more than twice the amount of all the oil, natural gas, and coal deposits combined, everywhere on Earth.

Recent mapping by the U.S. Geological Service, undertaken on the edges of the continental shelf off South Carolina, have shown a pair of hydrate accumulations, each about the size of the state of Rhode Island, each containing somewhere around 1,306 trillion cubic feet of methane—each enough to substitute for the natural gas consumption of the entire United States for 70 years. This small reservoir—many more are much bigger—contains maybe 35 gigatons of hydrate; all the other fossil fuels combined, everywhere, total only about 5,000 gigatons. There may be much more than that: not all the methane compresses into hydrates, just the top layer. Under the surface, temperatures increase, and so there are very likely massive amounts of methane deep in the sediment that never freezes. The frozen layer above acts like a lid to trap it in place.

Hydrates represent another peril: they are known to "cement" loose sediments in a surface layer several hundred feet thick and are sometimes the only thing holding the continental shelves together. If you took them away, or they depressurized or melted, massive landslides would follow.

By "mining" the methane hydrates very carefully, and by using the methane as fuel, we might be able to head off its sudden, catastrophic eruption into the air. Better to use it up, actually: if the hydrates were ever to dissipate, whether through rising temperatures, a lessening of pressure, or drilling through the hydrates into the reservoirs of gas below, and that methane was released into the atmosphere, calamitous global warming would inevitably ensue. There is, after all, somewhere around three thousand times the volume of methane trapped in hydrates as there is currently in the global atmosphere, and methane is a far more potent greenhouse gas than carbon dioxide, by a factor of ten.

Pretty good evidence exists that this has happened at least once before. Toward the end of the Paleocene, 55 million years ago, the Earth suddenly and rapidly warmed up (it was this warming, and not the end of the dinosaurs, that led to the rapid proliferation of mammalian species, eventually including our own). Temperatures spiked upward by 41 to 45 degrees Fahrenheit, as the oxygen isotopes stored in fossil plants clearly show, and persisted at that level for about 10,000 years. This upward pulse is called the Late Paleocene Thermal Maximum and was likely caused by the release of 2,000 billion tons of methane into the air. It is also called the Methane Burp theory.

The theory of the burp was given a substantial boost in 1999 by a research team from Rutgers University in New Jersey, led by paleoceanographer Miriam Katz. Her team found a sequence of sediment layers buried third of a mile below the sea floor off Florida, recording the exact sequence of deep-sea changes, including vast submarine landslides, that the theory predicted. As a report on her discoveries in *Science* put it, "If not for the evolutionary turning point [this] methane burp might explain, we might be eating egg-laying mammals for Thanksgiving—if we had evolved at all."

The theory was essentially this: at the time, somewhere around fifteen trillion tons of methane hydrate had formed beneath the sea floor, most of it by Archaea digesting organic matter. Then, for a reason still unknown, a small fraction of the hydrate, maybe a trillion tons, decomposed into water and methane gas. The methane, in turn, disrupted the sediment, triggering landslides down the continental slope that let a massive methane burst into the ocean On its way up, possibly, the methane oxidized to form carbon dioxide, which eventually reached the atmosphere, driving precipitous greenhouse warming.

More recently, German researchers reported that the same core Katz studied contains a time marker in the form of rapid, periodic variations in iron content. Their conclusions? Two-thirds of the trillion tons of methane was released in a few thousand years or fewer, about as fast as humans are releasing carbon dioxide by burning fossil fuel. "No one has figured out how that much carbon could have found its way into the environment so quickly, 55 million years before cars and power plants, except by destabilizing methane hydrates."[3]

In fact, smaller "burps" are not so uncommon. In the local lore of Maritime Canada, so-called mist-puffers occasionally burst to the surface, sometimes causing staccato explosions. In the early nineteenth century just such "puffers" so startled the villages of Passamoquoddy Bay (an inlet of the Bay of Fundy, between Maine and New Brunswick) that they called out the militia, thinking they were being attacked by the newly independent Americans. Virtually every bay along the Maine and New Brunswick coasts shows evidence of subsea pockmarks (small craters), hallmarks of methane bursts.[4]

Recent thinking about hydrates has focused on a more benign possibility, that the methane hydrates form one-half of a planetary feedback cycle, keeping the climate (at least over the long haul) relatively stable. The hydrates are kept solid by the pressure of the ocean above them. But in ice ages, the glaciation locks up so much water that the ocean levels drop, by more than 325 feet. This drop lessens the pressure on the hydrates and lets the methane bubble to the surface, setting off a greenhouse effect that in turn reverses the glaciation, bringing the planet back to its former state.

What's worrying, though, is this: "The natural controls on hydrates and their impacts on the environment are very poorly understood," according to William Dillon of the U.S. Geological Service, but a world greedy for energy and facing the ultimate end of cheap oil might blunder into exploiting it anyway, with uncertain consequences.[5] A report by Dillon reminded the world,

> Methane bound in hydrates amounts to approximately 3,000 times the volume of methane in the atmosphere. There is insufficient information to judge what geological processes might most affect the stability of hydrates in sediments and the possible release of methane into the atmosphere. Methane released as a result of landslides caused by a sealevel fall would warm the Earth, as would methane released from gas hydrates in Arctic sediments as they become warmed during a sea-level rise. This global warming might counteract cooling trends and thereby stabilize climatic fluctuation, or it could exacerbate climatic warming and thereby destabilize the climate.

But the prospects are enticing enough that some drilling will almost certainly be done. The "Carolina Trough" is particularly tempting. This is a very large basin about the size of the state of South Carolina, to which it is adjacent, with very deep sediments—by some measures, more than 8 miles thick. And because the methane in this sediment is held in crystals, it is more densely packed than in conventional sources; in addition, hydrates may form the "lid" on reservoirs of deeper free gas.

As Dillon put it, "These traps provide potential resources, but they can also represent hazards to drilling, and therefore must be well understood. Production of gas from hydrate-sealed traps may be an easy way to extract hydrate gas because the reduction of pressure caused by production can initiate a breakdown of hydrates and a recharging of the trap with gas." This could, however, also set off a runaway dissolution of the hydrates themselves, with potentially catastrophic consequences. Just such a scenario was the basis for the wildly best-selling German novel *The Swarm*, which postulated, in this case, an inimical undersea intelligence deliberately melting the hydrates to cause mayhem among land dwellers. Stipulation about intelligence notwithstanding, the actual science in the novel was good enough to have it reviewed by an oceanographer in the journal *Nature*—particularly the part where a collapsing hydrate formation in the North Sea caused massive tsunamis to sweep over low-lying countries adjoining. Alas, it seemed all too plausible.

ELEVEN

Tsunamis

He staggered and fell, grasped vainly at the stone, and slid into the abyss.
Fly, you fools! he cried, and was gone . . .
—J.R.R. Tolkien, *The Fellowship of the Ring*

Somehow this admonition—*Fly, you fools!*—kept coming into my mind as I read the warning bulletins and information statements issued by oceanographers and tsunami specialists everywhere, including my own small part of the world, the south shore of Nova Scotia. Because that seemed to be the sum of the advice these experts were able to give, buried in the verbiage about inertial moments, wave train lengths, earthquake seabed displacements, bathymetry, and the rest—when a tsunami is coming, to *fly!* is the only thing you can possibly do.

But how do you know when, or even whether, it is coming? The dismaying thing about tsunamis is that they render coastal cities vulnerable to unseen and unheard events that might take place halfway around the world, and they can come up on unsuspecting communities with terrifying speed—deepwater tsunamis have been clocked traveling at 584 miles an hour, faster than a commercial jetliner. Sure,

I was told reassuringly in 2006 that "Canada is updating and improving its tsunami warning system for the east coast with better earthquake sensors, coordinated through [the Oceanographic Institute at] Dartmouth, Nova Scotia"—this despite the historic evidence that the last tsunami to hit eastern Canada was in 1929, when lives were lost and villages destroyed in the Burin Peninsula, Newfoundland, and some damage done in Cape Breton, Nova Scotia. These warning systems consist of what, exactly? Bulletins issued to the media? What are we to do, live our lives with an ear glued to the nearest radio station, just in case? Even in the middle of the night? After which we would have—what, ten minutes? to *fly, you fools!*

In January 2005 the U.S. government, which has much more money than Canada's, made a $37-million contribution to an international consortium called GEOSS (Global Earth Observation System), which is developing a globally comprehensive, sustained, and integrated watch on what goes on around the world. The $37 million would set up thirty-two more deep-ocean buoys, mostly aimed at giving the United States "nearly 100 percent" detection capability for a coastal tsunami.

Such monitoring devices, albeit an older technology and not as widely deployed, have existed in the Pacific since 1946. This system monitored earthquake activity—the preeminent cause of tsunamis— and measured the passage of tsunami waves at tide gages. But while these measurements give a good indication of tsunami probabilities, they give no real information about the tsunami itself, or what impact it would have on specific pieces of coast. In fact, fifteen of the first twenty warnings issued turned out to be false alarms.

It has gotten better since. Measurements taken during the Rat Island, Alaska, earthquake in 2003, a magnitude 7.8 shaking, generated a tsunami that was detected by three of the new "tsunameters" placed in the Aleutian Trench and produced a real-time and very accurate forecast well before the waves reached the coast, correctly predicting amplitudes, arrival times, and periods of several of the first waves of the multi-wave event. These predictions would give some hours of warning to, say, Hawaii, but still only minutes to the Alaskan or British Columbian coasts.

Even so, predicting when and where the next tsunami will occur is currently impossible.

What, then, is the point? To encourage populations to live in a permanent state of high anxiety?

Well, it turns out there is a point. And part of that point is an educated, though hopefully not paranoid, population.

The USGS has analyzed the impacts of two different Asian tsunamis, one in the Sea of Japan in 1993, the other in Papua New Guinea in 1998. The Japanese tsunami killed about 15 percent of the at-risk population; in Papua New Guinea, about 40 percent died. The difference was attributed to risk-abatement strategies: the Japanese were educated about tsunamis and what to watch for and had developed evacuation plans and early warning alerts. Papua New Guinea, by contrast, had none of these things.

And in 2007, a mere fifteen minutes after a magnitude 8.4 earthquake rocked Sumatra, tsunami warnings were blaring from mosque loudspeakers in villages all along the coast—though by then the villagers, recognizing the early signs, had already fled. Vigilance works.

You could still die. But if you're vigilant, and you know where you should go, and if there is an uphill somewhere nearby, you could, indeed, *fly!*, and survive.

Tsunamis (from the Japanese "tsu," meaning *harbor*, and "nami," meaning *wave*) are more common, and much more destructive, than you might think. When I last dipped into NOAA's (National Oceanic and Atmospheric Administration) west coast and Alaska tsunami warning website, I found that 16 tsunami warnings (or "information statements") had been issued just for the previous month, 7 of them off Alaskan waters; 2 in the Molucca Sea, Indonesia; and 7 off the Kuril Islands, Russia.

Somewhere in the world, there is a significant earthquake causing a tsunami at least once a year on average, "significant" defined as causing substantial damage and loss of life. In the 10 years between 1996 and 2006, indeed, there were 10 such destructive events.[1] Global catalogs suggest that 157 tsunamis occurred from 1983 to 2001. Of these, all but 19 were in the Pacific region, clustered mostly around either Indonesia or Alaska. The 19 exceptions consist of 2 in the Indian Ocean, 9 in the Mediterranean, 1 in the Gulf of Aqaba off the Red Sea, 1 at Hanian Island in the South China Sea, 1 in the Marmara Sea (Bay of Izmit), and 5 in the Caribbean. Thirty of the 157 caused fatalities.

Even the little Mediterranean has recorded more than 300 tsunamis in the last 3,300 years, the most recent in December 2003, when a chunk of the Stromboli volcano slid into the Aeolian Sea, creating a 33-foot tsunami.

Tsunamis rank high on the scale of natural disasters, considerably higher than volcanoes. Since 1850, they have killed more than 420,000 people. Four of the deadliest in known history took place around Japan; other parts of Asia added three more. By far the worst in terms of its death toll was the earthquake and tsunami that struck Indonesia, Sumatra, and a cluster of Indian Ocean countries the day after Christmas in 2004. At time of writing, nearly four years after the event, the toll was put at 300,000 dead with thousands of people still listed as missing, and the relief effort had not come even close to winding down.[2]

As reported in the chapter on earthquakes, the event that set it off was massive, a 932-mile-long rupture of the Sunda "subduction megathrust" that caused a sudden lurch of the crust of up to 66 feet, at depths shallower than 12 miles, setting off an earthquake of between 9.2 and 9.3, the second largest ever recorded on a seismograph, and the largest since the Prince William Sound quake of 1964. It set off a tsunami that was recorded on tide gages worldwide (even in faraway Halifax, Nova Scotia, where a 20-inch surge showed up a full day later); the lurch was sufficiently powerful to change the Earth's rotation by a fraction of a second.

How big were the tsunami waves? Most photographs and video we have were taken by tourists at resort hotels, and while they are both fascinating and horrifying ˉbecause they plainly show how innocent most of the people were of the disaster about to enfold them, the waves don't look that massive, somewhere around 6.5 feet to maybe 13 or 16. This is misleading—the pictures come from survivors, and survivors were generally in stretches of coast where the topography minimized the damage. After-the-fact measurements, using laser rangefinders as well as more mundane markers like water stains on surviving buildings and broken branches and debris in trees, indicate that the west-facing coastlines were struck by waves more than 100 feet high, about the size of a 9-story building. Waves that hit the north-facing coasts, such as those of Banda Aceh, were lower, but still enormous, between 33 and 39 feet, and penetrated farther inland because of the low-lying countryside. The maximum runup, to use the oceanographers' term, in Sri Lanka was about 33 feet, but mostly lower.

Technically, a tsunami is any set of ocean waves (tsunamis never come in singles, always in sets) caused by any large, abrupt disturbance of the sea surface. The popular notion that a tsunami is a large breaking wave (like the classical Japanese Hokusai painting) is wrong: it is generally defined as "any sudden, non-meteorologically-induced impulse in water, regardless of size." Most tsunamis are set off by substantial earthquakes greater than 7 on the Richter scale, though not all—there is, in fact, a rather unnerving list of possible causes, which we will come to in due course. Among them are volcanoes (Krakatoa the most obvious example) and asteroid impacts (in 1997, scientists discovered evidence of a 2.5-mile asteroid that landed in the Pacific off Chile about two million years ago, setting off a huge tsunami that swept over portions of South America and Antarctica). But earthquakes, and the landslides they cause, are the dominant cause. As a consequence, most tsunamis, as you would expect, are set in train in earthquake-prone areas—the Pacific's Ring of Fire, the Mediterranean and Aegean, the Caribbean and a few other places. Most frequently they occur in the western Pacific, where dense oceanic plates are sliding under lighter continental ones, which fracture and can collapse.

Tsunamis may start locally, but they don't always stay local—see the 20-inch wavelet in Halifax, which seems to have traveled 1,500 miles to get there. They can cause local calamity within minutes, but a large tsunami can cause both local devastation and destruction thousands of miles away in distance and hours away in time, in which case they are referred to as teletsunamis. Because each earthquake is unique, and because the geometry of the coastlines differ so much, every tsunami has a unique signature of wave length, wave height, directionality, and speed.

Tsunamis almost always have a considerable wave length, sometimes so long that eyewitnesses will swear that only one wave crest arrived, since the next can be half an hour or more away. A tsunami arriving on shore will, typically, first "suck" the water out, like a high-speed low tide, a sure sign that fleeing immediately is the best, and probably only, option—if you see mud where a few minutes earlier there had been water, run for higher ground. When the wave does appear, it can pour in for fifteen to thirty minutes and sometimes longer; then the same water will rush out, often dragging hapless victims with it, for another thirty minutes or so, before the next wave arrives, as it surely will. The first wave is usually not the highest, either. It can be the third, or even fourth, that reaches the peak height. Some tsunamis have been known to contain no fewer than nine such runups.

A tsunami can carry almost a tenth of the energy of the event that caused it, a potentially huge amount. But because the wave length is so long, this stored energy usually results in only a small wave in mid-ocean, albeit one traveling at tremendous speed. Ships at sea often don't even notice if a tsunami passes beneath them. Tsunamis are really only destructive when they reach shallow water, when they slow down and are compressed, causing them to rear up, concentrating their accumulated energy in towering piles of water. In 1896 a tsunami raced into Japan, passing unnoticed beneath some fishing fleets, which, when they returned home, found their villages destroyed and their homes pulverized along nearly 300 miles of coastline (the official census tallied 26,975 dead, 5,390 injured, 9,313 houses destroyed, and 300 large ships stranded).[3]

Every now and then, someone survives a tsunami. A Catholic church in Banda Aceh made it intact through the 2004 tsunami, the few terrified parishioners who had taken refuge behind the high altar emerging to find their town utterly wrecked; a little down the coast, a man and a small child were found clinging to the high branches of a brawny tree after the waters had receded, a very unlikely refuge. (As America's NOAA put it on one of their tsunami web pages, "Going at once to high ground is the only safe escape

The Speed at Which Tsunamis Travel

The speed at which tsunamis travel depends on wave length and the depth of water. For a tsunami at a depth of 23,000 feet with a length of 175 miles, a not unlikely combination, the speeds will be these:

23,000 feet depth, wave length 175 miles, speed
   586 miles an hour;
13,000 feet depth, wave length 132, speed 443 m/h;
6,500 feet depth, wave length 94, speed 313 m/h;
650 feet depth, wave length 30, speed 93 m/h;
165 feet depth, wave length 14, speed 49 m/h;
33 feet depth, wave length 6.6, speed 22 m/h.

The shallower the water, the shorter the wave length, the slower the speed, the taller the wave—but the same amount of energy.[4] Just before landfall, tsunamis become mountains of water before which buildings, machines, forests and, of course, humans, are utterly helpless.

tactic, and waiting there until it is really over. Otherwise try roofs of sturdy buildings, or tall sturdy trees, but these are desperate acts.") Anecdotes of tsunami anomalies can be found from many places. In 1495, a tsunami of unknown cause struck the southern coast of Honshu Island, Japan, washing away a shrine containing an 36-foot bronze statue of Buddha weighing more than 800 tons— the statue was eventually found 165 feet above sea level, a mile away. In 1960 Chile was struck by the largest earthquake ever measured, 9.5 on the Richter scale, a shock that devastated more than 600 miles of coastline and killed more than 2,000 people, many of them from the series of 9 tsunami waves that followed (61 of the dead were in Hawaii and 122 in faraway Japan). But even here there were anomalies: in the town of Queule, about a mile from the shore, many houses were washed away. In one of them was a woman named Margarita Liempí; she stayed in her house while the whole structure moved for a mile, intact and upright—even her drinking glasses were unbroken. Another survivor, Filberto Henríquez, saw other houses floating away, with their stoves still smoking, looking for all the world like blunt and square-prowed ships. From the height of debris left tangled in the few remaining trees, Wolfgang Weischet, then a geographer at the Universidad Austral de Chile in nearby Valdivia, estimated that water was 13 feet deep in Queule.[5]

An earlier Chilean earthquake and tsunami, in 1868, occurred while the American civil-war era side-wheel steamer *Wateree*, 974 tons, was moored in the harbor of the town of Arica. The vessel and its 235 crew survived, but only in the most dramatic fashion: the boat quite literally surfed the tsunami crest and ended up almost half a mile inland and about 13 feet above what had been the high-water mark—so far inland was it, and so badly damaged, that the U.S. Navy sold it to an enterprising entrepreneur, one William Parker, who converted it to living spaces for the stunned refugees of Arica and later turned it into an inn. What saved the ship was its flat bottom; when the sea first receded, in the typical tsunami pattern, it grounded everything in the harbor, but the *Wateree* remained upright and floated off when the waves returned. Two other vessels in the harbor at the time, the Chilean gunboat *America* and the English bark *Chanarcillo*, had more normal keels, went over on their beam ends as the water rushed out, and were swamped and wrecked when the waves returned.

When the United States flagship *Powhatan* visited Arica a few weeks later, its commander, Rear Admiral Turner, described in a memo to the Secretary of the Navy what he saw:

> The upper part of the city, which from its elevation escaped the encroachment of the sea, has not a single house or wall left standing [presumably from the earthquake]—it is in one confused mass of ruins, more or less in every part prostrate; whilst the lower part, which comprised chiefly the better and more substantial order of edifices, including a large custom-house of stone mason work, is literally and perfectly swept away, even the foundations, as though they had never existed, and present the appearance of a waste that had been ravaged by the waters of a mighty river, carrying everything before it in its irresistible volume.

The *Wateree*'s own commander, James Gillis, was of course an eye-witness to the events. In his official report of the incident, he wrote:

> At 5:05 P.M. on that day a rumbling noise, accompanied by a tremu-lous motion of the ship, was observed. This increased in force rapidly until it was evident that an unusually severe shock of an earthquake was taking place, and I proceeded on deck, and, while standing there, looking at the city, I observed the buildings commence to crumble down, and in less than a minute the whole city was but a mass of ruins, scarcely a house being left standing.

The earthquake and its resulting tsunami claimed 25,000 lives. At Iquique, 120 miles to Arica's south, the receding wave uncovered the bay to a depth of 24 feet and returned with a 40-foot wave crest that engulfed the city. The tsunami was recorded at the Sandwich Islands, 6,400 miles away, only 12 hours and 27 minutes later, which meant it traveled at somewhat better than 500 miles an hour.[6]

Those of us living along the east coast of America have been, per-haps understandably and certainly unjustifiably, smug about the prospect of earthquakes and tsunamis. Sure, there was the 1755 Queen Anne's earthquake, the New Madrid earthquake of 1811, the Charleston earthquake of 1886, and the Grand Banks shaking of 1929, but that's about it. After all, we are nowhere near a tectonic subduction zone. Tsunamis must therefore be rarer still. No real vol-canoes, either. Yes, we have had hurricanes, but other natural hazards are scarce. Alas, this turns out to be spectacularly wrong.

That's because earthquakes aren't the only cause of tsunamis. Scientific thinking began to change after the 1998 Papua New Guinea earthquake, a moderate magnitude 7.1 event that set off a 50-foot tsunami, much larger than expected. Subsequent investiga-tions fingered an earthquake-triggered landslide as the probable cause. Still an earthquake, of course, if only indirectly, but oceanog-raphers began to examine other tsunamis with this in mind.

Tsunamis turned out to be less rare in the western Atlantic than commonly believed. Among the pieces of evidence are the massive coral-block erratic boulders, some of them the size of large ocean-going yachts, that are found on Eleuthera and in other places in the Bahamas. They have been the subject of intense curiosity for years and are now thought to have been pushed a third of a mile inland and 65 feet upward by massive teletsunamis coming in from across the Atlantic, after an event whose nature is still uncertain, but was possibly a landslide in the Canary Islands. Even in modern times, tsunamis are, if not routine, at least not so rare; researchers have counted about 40 in the region since 1600. In 1918, 216 people were killed when a tsunami with a runup height of 20 feet struck the western and northern coasts of Puerto Rico, after a magnitude 7.2 earthquake occurred on a fault line called the Mona Canyon, between Puerto Rico and Hispaniola.

The best known, and most destructive, tsunami to have hit the western Atlantic was the one set off by the Lisbon earthquake in 1755. Apart from the devastation in the eastern Atlantic to European and Moroccan shores, considerable damage was reported in Cuba and Martinique; 10-foot tsunami waves were thought to have battered the eastern seaboard of the United States.

The Burin Peninsula tsunami was locally generated, but not insignificant. The earthquake that set it off was only magnitude 7.2, but it created a 23-foot wave (up to 33 feet in places) that killed 28 people on shore and was recorded as far south as Charleston and as far east as the Azores. It was first thought to have caused a turbidity eddy that severed the transatlantic telegraph cables, but subsequent investigations, including submersible observations from the underwater explorer *Alvin*, have concluded that it was an earthquake-caused landslide on the continental shelf that set everything off.[7]

Cables have broken in the Atlantic before and since; that many of them occur after small earthquakes on the continental slopes suggests that slumping and landslides are fairly frequent. Any such landslide could set off a tsunami.

Here's the really scary bit: the prospect of a collapse or landslide in the Canary Islands, across the Atlantic off the coast of Morocco. No matter that they've been called the Happy Islands since Roman times, or sometimes the "Garden of the Hesperides," or that Christopher

Columbus was reported as liking the place when he stopped by on his way to whatever it was he found—the islands are volcanic in origin, and many of the volcanoes are still active. In December 2006 the journal *Science* reported on the work of Maria Pareschi, of Italy's National Institute of Geology and Volcanology—the same researcher who traced tsunamis resulting from collapses of Mount Etna and Stromboli, among others. "A massive collapse of Cumbre Vieja, a volcano in the Canary Islands, would trigger a towering tsunami that would pummel both sides of the Atlantic," Pareschi suggested. "Such a collapse would be 10 times larger than the Etna slide, [itself] an immense geological event. It would overwhelm New York, Miami and Lisbon."[8]

Miami, New York, Boston—and me, here in little Port Medway, Nova Scotia.

How tall would such a towering tsunami be? Possibly as high as 330 feet. That's as high as many of the buildings on Wall Street. Nothing on the coast would escape.

But why would Cumbre Vieja, which is on the island of La Palma, collapse? Because it is a volcano, it is active—and on its slopes, rather precariously perched, is a massive slab of rock the size of a small island (an "aggregated mass" in the jargon of geologists) that could easily be nudged off to plunge into the sea, causing the massive tsunami Pareschi has suggested. Aggregated masses, logically enough, cause greater tsunamis than rocks that fall as small pieces, or disaggregated, as they tend to do in Hawaii. Previous collapses have happened in the Canaries. There is already a large natural amphitheater—evidence of an earlier collapse—on La Palma itself, and others on other islands in the seven-island chain. The Bahamian coral blocks could have been thrown up by the tsunami this collapse set off; the ages match. A report in the *Journal of Volcanology and Geothermal Research* used ocean-floor imagery off the island of El Hierro to delineate a 24-cubic-mile landslide from a volcano called El Golfo, containing several blocks 0.6 mile across, and to its south, another younger volcano that appeared to be ready to collapse "at any time."[9]

In 2007 the BBC reported on a conference convened by several academic institutions to discuss global geophysical events and quoted a professor, Bill McGuire, of Britain's Benfield Grieg Hazard Research

Centre, as saying that "eventually, the whole rock will collapse into the water, and the collapse will devastate the Atlantic margin. We need to be out there now looking at when an eruption is likely to happen . . . otherwise there will be no time to evacuate major cities."

Not everyone agrees that calamity is imminent. Two scientists from the Southampton Oceanography Centre in England, R.B. Wynn and D.G. Masson, admit that Cumbre Vieja is unstable and concede that a major landslide could be catastrophic, but they suggest that the ocean floor shows several separate levels of turbidity sediment, indicating that prior slides were episodic and thus smaller. Consequently, they conclude, "the potential tsunami hazard from such failures may be lower than previously thought."

W ell, if a Canary volcano doesn't get you, perhaps a methane blow will.

We've seen from Chapter 10 how compressed undersea methane hydrates can explosively disintegrate, sending massive infusions of greenhouse gases into the atmosphere and setting off a series of equally massive tsunamis. These hydrate deposits are attracting the attention of some very heavyweight commercial interests, since they also imply a medium-term solution of the energy crisis.

In the Atlantic, broadly defined, the biggest pressurized methane hydrate deposits are in the North Sea east of Scotland and up past Denmark and Norway; off Cape Hatteras, along the continental shelf adjacent to North Carolina and Virginia; and off the coast of New Jersey. The North Sea deposits seem stable, but both the others are vulnerable to landslides. Neal Driscoll, of the Woods Hole Oceanographic Institution's Geology and Geophysics Department, along with a few colleagues, published a paper in 2000 that tracked newly discovered cracks along the edge of the continental shelf that he suggested could be an early warning sign that the sea floor is unstable.[10] If the sea floor slumps, it would trigger tsunamis—not 330 feet, considering the relatively smaller size of the slumping rocks and the shorter distances, but big enough, perhaps 33 feet or so. The cracks they discovered, called "*en echelon* cracks" (cracks that are parallel and overlap, rather like tiles on a roof) were found along a 25-mile section of the outer continental shelf, at water depths of 330 to 660 feet, and occurred somewhere around 16,000 to 18,000 years ago.

"[This section of] the outer continental shelf . . . might be in the initial stages of large-scale slope failure," Driscoll wrote. Already some of it has slipped 165 feet, in areas of large deposits of methane hydrate and pressurized water. In fact, the cracks may have been produced by methane bursts.

A second vulnerable section is off New Jersey, where Driscoll's team and other scientists have been exploring a series of largely unknown submarine canyons about 95 miles offshore.

But what would trigger the methane deposits to explode, thus triggering in turn the landslides and tsunamis? Warmer water would do it. So would reduced pressure. In the long term, global warming would be enough to raise the water temperature sufficiently. More worryingly, a major hurricane passing overhead could lower the surface pressure enough to destabilize the methane, which would then set off landslides and their consequent tsunamis—which would be larger than otherwise, added to the storm surge already caused by the hurricane itself.[11]

Hurricanes may cause tsunamis without the outside help of a landslide. As geophysicist Patricia Lockridge points out, "A number of East Coast wave events could not be directly associated with earthquakes. Some events were not clearly linked to any cause. Other events such as September 3, 1821, September 22, 1909, August 19, 1931, September 21, 1938, and September 14–15, 1944 occurred in association with hurricanes."

These "hurricane waves" are rare but not unknown. The preeminent hurricane forecaster Gordon Dunn, with his colleague B.I. Miller, once pointed out that "there is one additional and very important (though rare) sea effect which may occasionally be superimposed upon the usual hurricane tide—usually with disastrous results. This is the hurricane wave, sometimes erroneously called a tidal wave, although it has nothing to do with tides. It is nearly always described as a wall of water advancing with great rapidity upon the coast line."[12]

Whether that wall of water is a tsunami, a storm surge, a rogue wave, or a combination of all, coastal communities everywhere are at risk. No need to *fly!* just yet, however.

TWELVE

Floods

And, behold, I, even I, do bring a flood of waters upon the Earth, to destroy all flesh, wherein is the breath of life, from under heaven; and every thing that is in the Earth shall die.
—God, Genesis 6:17

Thus spake Jehovah, in one of his fouler moods, and ever since a bizarre amount of intellectual capital has been expended in finding the final resting place of Noah's Ark, into which the patriarch had shoehorned "every beast, every creeping thing, and every fowl, and whatsoever creepeth upon the Earth, after their kinds," for 150 claustrophobic and no doubt malodorous days. Most of these earnest seekers after evidence went to Mount Ararat in Turkey, since that's where the Bible says Noah came to rest; but some went farther— I have a book in my library by Olive MacLeod, one of those genteel Victorian ladies who became such intrepid explorers, who found the ark itself on the shores of Lake Chad, in the southern Sahara.

Well, Noah makes a great yarn and comes along with a moral lesson most useful to priestly castes. God destroyed man because he "saw that the wickedness of man was great," but the priests could help there, keeping people on the straight and narrow for their own good. It seems an awkwardly long way around to destroy every-one—couldn't God have just done the deed directly? Why the flood? But memories of the great deluge, it seems, are preserved in legends and folk tales from many widely separated cultures.

Older by far than Genesis is the epic of Gilgamesh, the Babylonian ruler who set off to find the secret of living forever. On his journey he met Utnapishtim, who had survived a great flood sent by the gods to punish wickedness and disobedience. Utnapishtim survived because he had been tipped off by the water god, Enki, to build a boat and so save his family and friends, along with artisans, animals, and precious metals. If the writers of Genesis copied this old yarn, they edited out the precious metals thing, which ill fitted the moral of the story they were telling; still, the tale would have been familiar enough to Moses and his contemporaries.

The great cultures of the old Mediterranean world, the Greeks and then the Romans, told a similar story. In this case Deucalion and Pyrrha had saved their own children and a clutch of beasts by bob-bing above the flood in a giant box. And to the west in Ireland, Queen Cesair and her retinue had taken to the sea in ships and stayed there for seven years when their land disappeared under the sea. Missionaries in Africa and the Americas collected dozens of similar tales, and being Bible-centric, they often suspected the leg-ends had been planted there by earlier travelers, or survivors, or God himself, or possibly God's own rival, Satan. For example, the Bozo people of Mali and Niger, river dwellers and boatbuilders, have a legend that recounts a great battle between Water and Land, in which Water first prevailed but finally lost to Land, was imprisoned in the Niger River, and the Bozo became its custodians. Surviving the Great Deluge was a common theme, too, in aboriginal American legends.

In 1999, two Columbia University marine geologists, William Ryan and Walter Pitman, proposed a theory to explain all this. The Noah legend and its antecedents, they came to believe, were based, if rather loosely, on a real event—an event that dated to the very earliest days of human urban culture, in the aftermath of the last glacia-tion, somewhere between seven and eight thousand years ago. As the glaciers melted, the seas rose—and rose, and rose. It is a well-known geologic fact that the Ice Age had dropped sea levels by 30 feet or more, and the rise would have been steady and relentless. The Mediterranean, by then open to the world's oceans through what was known to the Greeks as the Pillars of Hercules, rose just as the oceans did. The Black Sea, however, remained a discrete body, still fresh and not salt, cut off from the oceans by a narrow section of Turkey. It has also been known for some time, from the evidence of mollusk fossils, that sometime around seven thousand years ago the Black Sea's water did change, rather abruptly, from fresh to salt. Pitman and Ryan speculated that the rising oceans had forced an abrupt break through what is now the Bosporus—essentially a dam breaking—and massive amounts of water cascaded through to the east, inundating settlements and causing thousands to flee. How much water? "Ten cubic miles of water poured through each day, two hundred times what flows over Niagara Falls . . . at full spate for at least three hundred days." More than 60,000 square miles of land were drowned, and the Black Sea expanded its banks, with catastrophic effects on the surrounding populations. The death toll, obviously, is unknown, but the event would have set off waves of refugees, fleeing to every known culture.

Their theory is not without its critics. The most vigorous of these was a team from Rensselaer Polytechnic headed by Jun Abrajano, who argued that the Black Sea rose gradually for two thousand years, starting at least ten thousand years ago, hardly the catastrophe that the theory suggested.

But the Ryan-Pitman hypothesis received a substantial boost when maritime explorer Robert Ballard, the man who found the *Titanic*, was hired by the National Geographic Society and duly found complex human artifacts buried beneath the silt some 12 miles off the Turkish coast and about 325 feet down, at precisely the right level to have been seven or so thousands years old. "A stunning confirmation," Ryan was quoted as saying afterwards. " . . . [I]t shows unequivocally that the Black Sea flood took place. . . ."[1]

This flood might explain the Noah legend, but how about the American aboriginals? Different place, same idea: melting ice causing massive pileups of fresh water, abrupt breaks. . . . Maybe more than one break, in this case.

Considerable evidence exists that shortly after the end of the last ice age (starting at around twelve thousand years ago), cataclysmic floods inundated portions of the Pacific Northwest from Glacial Lake Missoula (in the Idaho Panhandle and Montana, where Lake Pend Oreille is now), pluvial Lake Bonneville, and perhaps from subglacial outbursts. Missoula, possibly as large as lakes Ontario and Erie combined, formed from glacial meltwater and was held back by a lobe of the Columbian ice sheet. Eventually, the water forced its way past the ice dam, inundating parts of the Pacific Northwest.

It was geologist J. Harlen Bretz who first proposed that the Columbia Basin in eastern Washington state had been scoured by a sudden cataclysmic flood. His notions were pooh-poohed by his colleagues for several decades before they finally accepted that features such as the Wallula Gap and the Drumheller Channels, both now U.S. national monuments, were most probably caused by flood scouring after some vast lake emptied, unleashing powerful erosive features. This version of Noah is now sometimes called the Missoula flood, or the Spokane flood, or even the Bretz flood, after its premier theorist.[2]

Something similar seems to have happened elsewhere in North America as the glaciers retreated. In the center, glacial Lake Agassiz, named after Louis Agassiz who pioneered glacial geology, covered an area five times the size of North Dakota today; but because the region was, and still is, flat, no massive flood breaks occurred. To the east, however, there is evidence of flood bursts from the "Champlain Sea," and into the Atlantic via the Champlain and Hudson river valleys,[3] and also down the modern St. Lawrence river valley and out into the gulf, pushing miles of sediments to the continental shelf.

Such events were not, obviously, confined to the Americas. Compelling evidence has emerged that Britain wasn't just gradually cut off from Europe, but that it happened in a single catastrophic flood event, when a rock dam in what is now the Dover Strait collapsed, sending floods of glacial meltwater racing northward to sever the last of the land bridges, a chalk ridge called the Weald-Antois Anticline.[4]

Floods are one of the natural phenomena most easily grasped. This is because while they can and do happen on a catastrophic scale (viz. Noah, above) they are also quotidian events, easily seen and often experienced, and happen as readily in tiny creeks as in huge rivers— indeed, they are really part of the same process, for a spring sometimes becomes a creek, which turns into a stream, which becomes a river . . . . We have a small creek on our property, usually not much more than one and a half feet wide at low water; after a heavy rain it readily doubles in size, and once after we received 4.7 inches of rain in a day and a half, it quadrupled and turned from a gurgle into a roar, overflowing our road and tearing away at tree roots and grasses. Scale this up to river size, and we've all seen the pictures—flood waters racing through settlements, overflowing bridges, washing away vehicles, battering away at riverbanks, and terrified householders clinging to rooftops.

Often, of course, they can't cling any longer, or the buildings themselves are destroyed. Then they drown. Many thousands of people have drowned, over the centuries of human history, in one flood or another. Millions, in fact—sometimes millions in a single catastrophic incident.

Water drives a hard bargain. You need it for life and to nourish your crops and animals. But it can take as well as give and can turn nasty in an instant.

Flooding has been the most life-destroying of all natural calamities.

One day in 2003 I stood on the bank of the Yellow River as it roared through a canyon in Shaanxi province, in north central China. I was there to film a television documentary on water, and the crew had wanted a grab shot for the B roll (which they called insurance and I called filler), and so I stood on a promontory staring down at the torrent below, while a crewman rather rudely shooed away Chinese honeymooners taking each other's pictures in front of the rapids. The honeymooners were there because this is, after all, one of the grand sights of China and one of the most famous rivers in that populous country: the Hwang-Ho, or Huang He, the Mother River of China, sometimes just called "the river," its central reaches nourishing the plains that were the cradle of Chinese civilization. It is the second-longest river in China after the Yangtze; it rises in the Bayankala Mountains in Qinghai province in the far west, flows through 9 populous provinces, and empties into the Bohai Sea. Every Chinese schoolchild knows the numbers: it is 3,395 miles long; its basin has an east–west distance of 1,181 miles and a north–south distance of 686 miles, for a total area of 290,520 square miles.

For my part, I was there because the river has another name: China's Sorrow.

The Huang He floods often, and it floods massively. Imperial scribes have accounted for 1,593 episodes of flooding in the last 4,000 years. The worst flood of all killed 4 million people in one miserable episode.

The water of the Yellow River is—this shouldn't be a surprise— yellow. It is yellow because it contains a vast amount of silt (if you were to scoop up a bucketful, as much as 60 percent of the contents by weight would be fine-grained yellow silt). The silt is picked up by the rapidly flowing waters pouring through deep canyons carved into north China's great loess plateau. "Loess" is defined by geologists as "an unstratified, usually buff to yellowish loamy deposit found in North America, Europe and Asia, and thought to be deposited by wind." China's loess plateau covers more than 250,000 square miles, about the size of Texas, and has been riven with great erosional gullies and canyons; millions of people still live in homes carved into the friable canyon walls, their chimneys often seen sticking bizarrely up through the planted fields above. The loess silt in the river, 1.6 billion new tons of sticky yellow mud every year, is one reason it has become so deadly. Over the years (though it has changed its course many times over the millennia) it has deposited thick layers of silt along the riverbed as it flows through central China's plains, with the curious effect of raising the riverbed itself, often to levels higher than the surrounding flat plains. Thousands of miles of dikes have been built over the centuries to contain the river, but in sharp floods they often give way (or are overtopped), with catastrophic results. The flatness of the plains means that every flood covers hundreds or thousands of square miles of heavily populated land.

It is hard to know whether to classify the first of the truly catastrophic Huang He floods as a "natural" disaster or not; it was the Yellow River that caused the mass deaths in Kaifeng, in east central Henan province, in 1642, but it wasn't accidental. The waters of the river were deliberately loosed by the Ming army (they cut through the dikes in several places) to prevent the peasant rebels under Li Zicheng from taking over the city. That did the trick all right, but as collateral damage, to use the current military euphemism, about half the city's population of 600,000 drowned or perished in the subsequent famine and plague, making it either the seventh-largest natural disaster of all time or one of the greatest criminal acts of war, genocide excepted, in the grisly history of human conflict. The city was essentially abandoned for 20 years afterwards and was rebuilt under Kangxi, the first of the succeeding Qing dynasty emperors.

The Yellow River flooded often during the next 200 years, on average once every 3 years, but the next real catastrophe wasn't until 1887, when heavy rains upstream caused the river to swell to nearly twice its size, flooding the North China Plain, causing somewhere between 900,000 and 2 million deaths.

Forty years later, in 1931, it happened again. More heavy rains, more flooding of the North China Plain, by now even more heavily populated. This time nearly 4 million (an estimated 3.7 million) people died, making it the single largest natural catastrophe ever. Nothing had been done in the interim to mitigate flood damage.

In 1938, during the war with Japan, the Nationalist troops under Chiang Kai-shek replicated the Ming cruelty by breaking the dike holding back the Yellow River, to stop the advancing Japanese army. More than 21,000 square miles of farmland were inundated, and more than half a million people died (estimates are between 500,000 and 900,000), more than were killed in the major battles of the war itself.

C hinese attempts to prevent flooding and control the river go back a long way—all the way to 300 BC, when a proto-engineer named Yu proposed dredging to control the silt and nudge the river back into its main course; afterwards, he became emperor for his pains, but the Yellow continued to flood, and in the centuries that followed thousands of miles of levees were built, channels dug, and dams erected. The most recent of these dams, first proposed by Chairman Mao and initiated later by Deng, is the Xiaolangdi Multipurpose Dam Project, 25 miles north of Luoyang City in Henan province, and is now China's second-largest dam. It was built concurrently with the largest dam of all, the infamous Three Gorges, but drew little of the same visceral opposition. Partly this was because the scenery was much blander, but it was also because even environmental groups with an instinctive opposition to dams of all kinds acknowledged that something had to be done to mitigate the flooding that gave the Yellow its dismal reputation.

Xiaolangdi isn't a permanent fix. After twenty years or so that relentless loess sediment will begin to clog its intakes and build up against the dam wall. "At that time," said Wang Xianru, deputy director for the contractors who built the dam, "our grandchildren will need to think of something [else]."

T he Yangtze (or Chang Jiang, Long River in Mandarin) is as prone to flooding as the Yellow, and partly because it is, indeed, long—the longest river in Asia and the third in the world after the Nile and the Amazon, stretching 3,964 miles from its source in northeast China's Qinghai province to its mouth at Shanghai in the southwest. The rainy season is May and June in areas south of the Yangtze, and July and August to its north, and so the wet season is effectively doubled every year. Human actions have exaggerated the effects; massive deforestation upstream has removed the reservoirs that once cushioned the flow, and flooding continued all through the twentieth century: 100,000 killed in 1911, 140,000 in 1931, 142,000 in 1935, 30,000 in 1954; and in 1998, a major catastrophe was narrowly averted when the banks overflowed and threatened retaining dikes. Had they broken, hundreds of thousands more would have died, and several major industrial cities would have been inundated. As it was, more than 2,000 people drowned, 13.8 million were driven from their homes, 2.9 million houses were destroyed, and 9 million hectares of crops ruined (3 percent of the national total); the floods affected an area inhabited by more than 240 million people, a fifth of China's considerable population.

At one point late in August, as the third flood peak of the summer poured through the not-yet-complete Three Gorges Dam and into the middle Yangtze, heading for the city of Shashi, the vice minister of water affairs, Zhou Wenzhi, suggested that if the water crested above the 148-foot danger point, it could be necessary to deliberately dynamite the dikes, flooding areas where half a million people lived, in order to save larger and more populous cities downstream.

China-watcher Lester Brown, head of the Worldwatch Institute in Washington, D.C., wrote shortly afterwards:

> The Chinese government is treating this disaster as an act of nature, and indeed it is. Floods during the monsoon season from June through September in southern China are a regular occurrence. But there is also a human hand in this year's floods in the form of defor-estation and intensive land development. The Yangtze basin is home to 400 million people, making it one of the most densely populated river basins on Earth. . . . With such a density of population, the human pressure on the land is everywhere. To begin with, the Yangtze river basin, which originates on the Tibetan Plateau, has lost 85 percent of its original forest cover. The forests that once absorbed and held huge quantities of monsoon rainfall, which could then percolate slowly into the ground, are now largely gone.

If the immense Three Gorges Dam had been built primarily as a flood control device, it would be a failure. Three different flood types happen along the river: in the upper reaches, in the lower reaches, and river-long floods; the dam can control only the third type. But in fact flood control was only one of three reasons for the dam's construction, and both the other two were regarded as more important. They were to generate hydroelectricity to offset coal burning; and to act as a reservoir that could be used to convey massive quantities of water northward, to the arid North China Plain, part of what is intended to be the single largest re-engineering of a national water supply in human history.

Flood waters sometimes collapse even the most robust dams. An egregious example was that of China's Banqioa dam, on the Ru River (Ru He), built in the 1950s, to generate electricity but with flood control as its main objective—severe floods had occurred in the Huia River basin several times in the 1940s and 1950s. The dam was supposed to be "over-engineered" to be safe—it would withstand a 1-in-1,000-year rainfall, over 11.8 inches a day. Alas, in 1975 Typhoon Nina collided with a cold front and dumped more than a year's rainfall in less than a day. At one point the rain was coming down in solid sheets, 7.4 inches an hour, over 39 in a single day, more than three times the extraordinary event the dam was built to withstand. When the wall finally burst, it sent 1.75 billion cubic feet of water plunging down the river. All in all, 62 dams burst in sequence, and a massive wave more than 6 miles wide and more than 20 feet high, with 15 billion tons of water behind it, roared down the valley at 37 miles an hour. An area 34 miles long and 9 wide was completely wiped clear of living creatures, vegetation, and buildings. A new lake 82,000 square miles appeared overnight. Seven county seats were inundated. Some 26,000 people died immediately, and another 145,000 in the days to follow. Nearly 6 million buildings collapsed. Eleven million residents were affected.

Chinese government reports acknowledge that this wasn't the only dam failure that catastrophic summer. In that year alone, dam collapses killed a quarter of a million people and brought famine and disease to more than eleven million others.

Nor are dam collapses exactly unknown elsewhere. In 1979, for example, the Morvi dam in Gujarat state, India, burst in a severe monsoon and sent a wall of water racing through the town of Morvi, killing, by some accounts, as many as twelve thousand people. In Uzbekistan in central Asia a catastrophic thaw in a glacier in mid-July 1998 produced flooding that destroyed a dam on the Kuban-Kel Lake, causing it to collapse; forty-three people were killed.

British Columbia's W.A.C. Bennett Dam, one of the world's largest earth-filled structures (it is 1.25 miles across the top and 600 feet tall; its Williston Reservoir covers 641 square miles and holds 70,308,930,000 tons of water), had some uneasy moments in 1997 when two large sinkholes were discovered just downstream. Emergency repairs were undertaken and in 1998 BC Hydro, which operates the dam, spent $4 million reinforcing its drainage system by gouging out a 16-foot-deep trench and backfilling it with rock. Ron Fernandes, then BC Hydro's area manager, assured everyone who would listen that the dam was perfectly safe. "Geophysical tests confirmed the success of the sinkhole repairs last year," he said, "and we plan on conducting these tests on an annual basis as part of our ongoing surveillance program."

Even in the United States, a country that prides itself on its engineering skills, dam collapses occur. The Teton Dam on the river of the same name collapsed in 1976, even before it was completely finished, wiping out three towns and hundreds of thousands of acres of farmland, which the flood scrubbed down to bare rock; fortunately there was enough warning that there were no casualties.

At the Glen Canyon Dam on the Colorado River in Arizona, a section of spillway was destroyed by flood waters in 1982. The spillway runs directly through the sandstone underlying the dam, and the writer Philip Fradkin described the event this way:

Two thousand tons of water per second soared from the spillways and the river outlet works on both sides of the dam in the most spectacular display of cascading water ever seen on the artificially controlled river. Rumbling noises were heard on June 6. The jets of water became erratic. Water the color of diluted blood, chunks of rocks, and pieces of concrete issued from the mouths of the spillway tunnels, as if the dam was mortally wounded.

The level of the reservoir peaked, Fradkin points out, at six-hundredths of a foot below the point where engineers forecast control might have been lost. Afterwards, the Bureau of Reclamation declared to its satisfaction that the dam itself was never in danger, but written documents at the time showed that bureau engineers "feared for the safety of the dam and its foundation."

More reprehensible was the Johnstown, Pennsylvania, flood of 1886. Record rains drenched the deforested Conemaugh River, and a dam that had been built and was maintained only for the benefit of a millionaires' fishing club (Andrew Carnegie and others) broke, turning a flood into a disaster with more than three thousand dead. It was reprehensible because the club had ignored repeated warnings that the dam was unsafe. After it was over, its assorted millionaires essentially bought a judge to rule the collapse an act of God, leaving the South Fork Hunting and Fishing Club with no legal liability. Most of the millionaires contributed only a token amount to the relief effort.

S ometimes dams, by their very presence, cause downstream problems by removing valuable sediments from the system. The High Aswan Dam in Egypt is a notorious example. Sediment loss has also been an American problem, and specifically a Mississippi River problem. In this case it has been a slow-motion disaster: the levees built to contain the mighty Mississippi flooding have ended up causing the loss of vast quantities of land to the sea; since the 1930s, an area the size of Delaware has been lost forever. But hardly anyone noticed— until Hurricanes Katrina and Rita roared through, finally revealing the dangers of the sinking coast and exposing what the Army Corps of Engineers had wrought over the years.

The Mississippi River begins as a tiny creek up in northern Minnesota, near the Canadian border, and flows 2,361 miles to the Gulf of Mexico in Louisiana, along the way drawing nourishment from many hundreds of tributaries in 31 U.S. states and two Canadian provinces. It is the sixth-largest river in the world in volume, with an annual average flow rate of 495 cubic feet per second and a freshwater discharge onto the continental shelf of 139 cubic miles a year. About two-thirds enters the sea in the bird's foot delta near New Orleans, the rest through the Atchafalaya River, a little to the west.

For millennia, presumably since the beginning of the current geological epoch, the Mississippi has overflowed its banks in high

summer rains, spreading over thousands of kilometers of low wetlands, whose natural sponge-like quality absorbed the surplus, releasing it in dryer weather. At the same time, the silt load of the river
helped sustain and extend the lands around the delta.

But then came people. People didn't like wetlands much.

They liked them even less than usual when the river flooded in
especially heavy rainfall years, such as 1927.

The Great 1927 Flood started in the summer of 1926, when torrential rains descended in the central Mississippi basin, in Kansas, Iowa, and
Tennessee. By September the river was near the top of its banks; one of
its tributaries, the Cumberland at Nashville, reached the top of its levee,
56 feet above normal water. Just after New Year's Day the Mississippi
itself broke out of the levee system that was protecting riparian cities
and spread out over 27,000 square miles in a wash 33 feet deep, killing
246 people in 7 states and causing what was thought to be
$400 million in damages. By late spring, the "river" below Memphis
was still 62 miles wide.

New Orleans awaited the flood with dread. To protect the city—
or at least the wealthier white areas—a dike at Caernarvon, Louisiana,
was blown up with 30 tons of dynamite, which sent 250,000 cubic
feet a second flowing through the mostly black St. Bernard and
Plaquemines parishes. Louisiana's (and Mississippi's) unabashedly
racist politics were exposed for all to see. More than 330,000 black
citizens of the southern states were moved to 154 relief camps; 13,000
of those, scooped up from farms surrounding Greenville, Mississippi,
were stranded on the crest of an unbroken levee without any food or
water and were forced to watch as a series of boats arrived to evacuate whatever whites remained. Other blacks were forced at gunpoint
to leave their families and homes and add their unpaid labor to the
relief effort. Conditions in the camps were so bad that Herbert
Hoover, the president, felt obliged to suppress the reports that
emerged, an action that forced a major switch of black voters to
Roosevelt's Democrats.

The main aftermath, though, was an instruction to the Army
Corps of Engineers to finally tame the river. They did this by
building the largest and longest levee system then known, adding

twenty-nine new locks and dams on major tributaries, and digging thousands of runoff canals. It worked, too, attracting literally millions of people to immigrate to the mid-Mississippi basin, where the soil was so remarkably fertile.

Or it worked until the floods of 1993. In that year, four out of five of the levees failed.

That summer, the whole basin received what the American Geophysical Union said was "anomalously high rainfall . . . [caused by] an abnormally persistent atmospheric weather pattern consisting of a quasi-stationary jet stream positioned over the central part of the nation; there, moist, unstable air flowing north from the Gulf of Mexico converged with unseasonably cool, dry air moving south from Canada." In fact, this anomalous pattern wasn't confined to the Mississippi basin but occurred over much of the northern hemisphere and was widely blamed on the existence of excess cloud-condensation nuclei from the ash cloud generated by the explosion of Mount Pinatubo, in the Philippines, two years earlier. The previous year, 1992, had also yielded heavy rains and also abnormally heavy snowfalls that winter.

Between April and August parts of the Midwest were deluged by up to 48 inches of rain, some 400 percent or more above the norm. Around St. Louis, in Missouri, all 36 measuring points recorded record levels in the region's rivers. In general, the 1993 flood exceeded levels set during any earlier floods. Discharge in August 1993, usually a low water month, exceeded that of an average month during the annual spring flood.

In the end, 324,326 square miles were inundated, causing $15 billion in damage, the most costly flooding in U.S. history.[5]

Until Katrina took down the levees in New Orleans.

Nevertheless, the straitjacket imposed on the Mississippi by the Corps of Engineers remains.

In May 2007, however, the Louisiana government, finally stung beyond endurance by reports of incompetence and malfeasance following Katrina, not all of which could be blamed on the hapless federal bureaucracy, FEMA, proposed rebuilding the delta entirely, recreating more than 100,000 square feet of wetland and bayou by, essentially, rerouting the river. The main point is to reclaim the sediment coming down from the fertile Midwest, which now flows—uselessly—into the sea. To do this, levees would be deliberately breached in a dozen places and new waterways created, to flow to the eroding areas, there to deposit their precious silt. The plan also proposed massive pumps to push silt to areas where marshes and barrier islands could be rebuilt. Most controversially, it would entail closing the shipping channel called the Mississippi River Gulf Outlet, which acted as a conduit for Katrina's storm surge and contributed substantially to the damage the hurricane caused. America's grain exports would have to go through locks, instead of steaming directly out to sea.

Whether it will work—or indeed whether the $50 billion or so it will need will ever be forthcoming—is as yet unknown. It will not be a quick fix, even if it does. "It could be hundreds, or even thousands, of years before we see a [new] spot of land," according to Kerry St. Pe, the director of a national estuary program. It will be too long: "Right now, we are at absolute collapse."[6]

Sometimes flooding takes place in localities that are so predictably prone to inundation that you wonder why anyone ever settled there in the first place. Such a place is the relatively new nation, Bangladesh.

The whole country is a low-lying delta to the northeast of the Indian subcontinent, in an area prone both to heavy monsoon rains and to typhoons. It is a country traversed by more than 250 rivers, some of them major arteries that start in the snow-covered mountains of India, Tibet, Nepal, and Bhutan. Among them are the Ganges, the Brahmaputra, the Teesta, and the Meghna; the Meghna alone swells to 5 miles wide during the seasonal rains. Even in a normal monsoon, floods cover more than 20 percent of the densely populated country; in severe monsoons the floods cover an astounding 60 percent or more—three-quarters in the serious floods of 1998, including two-thirds of the capital city of Dhaka, a city of more than 10 million people. More than 3,500 people died, and $2 billion worth of crops were destroyed. It happened again in 2004—another 1,000 people killed, another $2 billion in damage. In the last 100 years, regional flooding has killed more than 50,000, left 32 million homeless—about the population of Canada—and affected more than 300 million.

In 2007 it happened again, when the country was once again battered by mudslides, flooding, and ferocious thunderstorms; extra-heavy monsoon floods inundated a third of the country and left more than 5 million people stranded. Across Asia, the toll was even higher—almost 20 million people were displaced. Several hundred people were killed in India and scores missing as mudslides washed away hillsides near the city of Chittagong. In Pakistan, cyclone rains and flooding left perhaps 500 people dead or missing, 250,000 homeless, and more than 1.5 million affected in the southern provinces of Balochistan and Singh, setting off waves of protest against tardy relief efforts from the capital. China didn't escape and was once again reeling from widespread flooding.

The flooding is made worse by widespread deforestation upstream, which dramatically increases the flow rate of the water. Thousands of miles of embankments have been built since the 1950s, but they have mostly been colonized by the myriad poor and are easily broken. In the early 1990s a $10-billion plan to increase the scope of the embankments was scaled down by major donors to $5 billion, then abandoned altogether for a variety of ecological and engineering reasons. Most of the sluices and drainage outlets that exist are choked with garbage and plastic refuse, rather defeating their purpose. (The same thing happened in Mumbai, formerly Bombay, in 2005, when monsoons dumped three feet of rain in a mere 24 hours; the drains were choked with plastic bags and drained nothing, with the result that a quarter of the city was flooded, drowning 1,493 people in what the locals have come to call Terrible Tuesday.) The Bangladeshi government has attempted to get a regional conference together to deal with the problem on a multilateral basis, but with no success.

The upside of all this, if you can call it an upside, is that Bangladeshis are famously resilient in coping with flooding. A report for the International Food Policy Institute pointed out that even though two-thirds of the country was inundated in 1998, and more than two million tons of rice were destroyed (almost 10 percent of the annual production) there were relatively few flood-related deaths, and none due to food shortages. The 2001 report detailed, with some evident astonishment, how government policy, well-functioning private markets, household coping strategies, and donor and NGO interventions combined to avert a major food crisis.[7] Even the protective embankments that did exist came to serve a multiplicity of purposes, being used as highways, as a location for impromptu housing, as croplands and for grazing, as markets, as places to dig wells, and as emergency shelters. The silt deposited by the flood, while a contributing cause of ever-shallower riverbeds, was very fertile, and when the waters receded, new crops were at once planted; often these crops were sown in new islands of silt left after the monsoons departed, islands that were soon colonized by land-hungry residents from the slums of Dhaka, who quickly planted crops, becoming self-sustaining and self-sufficient in only a few months. Food grain output increased 16 percent in 1989, after serious floods that year, and by nearly 14 percent in 1999, after the widespread flooding a year before. The population has become more than just inured to flooding—they actually cope rather well.

As, indeed, do the Dutch, another improbable country built largely below sea level (three-quarters of the land area, and fifteen million people). The Dutch, in a nice piece of positive thinking, say that "God created the world in seven days, except for the Netherlands. The Dutch took that from the sea." It seems that the sea has been trying to get it back ever since. And, often, nearly succeeding.

The first known attempt to get the country back was the flood of December 838, in which the province of Friesland (that is, about 50 percent of the country, and almost all of the low-lying areas) was inundated, drowning 2,437 people. That we know the figure with such precision was due to a rather offhand note in the diary of the Bishop of Troyes, who called it the Eerste Grote Waterramp, or the First Great Water Disaster. Other disasters followed with depressing regularity: the Tweede Noodramp (second disaster) in

September 1014; the St. Martin's Flood of 1099; the Great Flood of 1212, which claimed more than 60,000 lives; the St. Marcellus flood of 1219 (36,000 lives); two floods on St. Elizabeth's Day (19 November 1404, and 19 November 1421); a flood attributed to St. Felix; and the All Saints Flood of 1570. The second of the St. Elizabeth floods and the All Saints one were caused by hurricane storm surges and resulted in enormous damage and substantial loss of life.

Dutch attempts to protect themselves and their country have been unrelenting. The very earliest settlers, driven by who knows what pressures to settle where settlement was imprudent, protected themselves as best they could by heaping earth up into mounds called terps, little artificial hillocks, on which they built their dwellings and farms. Gradually, over the centuries, they extended their dominions by constructing networks of dikes and draining the resulting generally quadrangular parcels of land, which came to be called polders. The famous Dutch windmills were put to use keeping these polders clear for planting; by the late Middle Ages there were windmills in nearly every field and on every dike; they simply became part of the landscape, as can plainly be seen from almost any Dutch landscape painting. So effective were they at keeping the marshes of the Rhine delta free of water that they turned Holland into an industrial powerhouse, a center of heavy industry and a major exporter, all powered by the wind—the oil of its time. The drainage system was financed by wealthy merchants from Den Hague and Amsterdam.

Larger-scale projects were mooted too. In the seventeenth century an engineer named Jan Leeghwater (ironically, the name means "low water" in Dutch) proposed reclaiming the Haarlemmermeer, or Haarlem lake, a 9.3 square mile tract between Amsterdam and Leiden. Nothing came of it for two more centuries, until the advent of steam pumps. In the meantime, there were more floods, including on the Haarlemmermeer itself. In 1836 two consecutive hurricanes, in November and on Christmas Day, explosively expanded the *meer* to nearly double its size, gobbling up villages as it did so, and sending waves of flood water into Leiden and almost to Amsterdam itself. Finally Leeghwater's dream was fulfilled: more than 37 miles of dikes surrounded the lake, and it was pumped dry by three massive steam engines, which sent more than 800 million tons of water back into the sea.

Another round of flooding on the Zuiderzee in 1916 promoted the building of the massive 18.6-mile Afsluitdijk (Closure Dike, or, as it is better known in English, the Barrier Dam), which had the effect of turning the Zuiderzee itself into a lake, the Ijsselmeer, and cutting the exposure to the sea from 186 miles to just 30. It was finished just before another kind of typhoon hit Holland, this one the Nazi onslaught.

The country was struggling to recover from the cataclysm of World War II when, in 1953, nature struck again. A combination of a very high spring tide and a major northwesterly gale system drove ashore a tidal surge that in places reared up to nearly 20 feet above sea level; it overwhelmed parts of the low-lying coastlines of England, Belgium, France, Denmark, and—of course—the Netherlands. This North Sea flood battered through the dikes protecting the vulnerable southwest of the country, flooding almost 773,000 square miles of land, destroying 300 farms, and sweeping away 1,835 people.

The storm replicated to an eerie degree another, the Burchardi Flood (sometimes called the Second Great Mankiller), that had struck about three hundred years earlier, in 1634, when another storm-driven surge overran dikes and caused thousands of casualties.

That there wasn't a much greater disaster was something of a fluke, a kind of modern twist to the old finger-in-the-dike legend. A great section of the country, home to three million people, was protected by a single massive dike, the Schieland High Sea Dike, parts of which hadn't been tested for centuries and at least one section of which was just beaten earth, with no stone reinforcement. And indeed, at 5:30 in the morning, a section of the dike crumbled and collapsed. It was plugged by the brilliant if unorthodox solution of commandeering a riverboat, the *Twee Gebroeders*, or Two Brothers, and driving it at full speed into the gap, where it wedged tight. The rest of the dike held.

The Water Management ministry had feared just such a catastrophe and, in an ironic coincidence of timing, had published a report less than a week earlier, urging the damming of every tidal inlet and estuary in south Holland. The floods that followed propelled the government into action, and the first barrier, a moveable storm surge barricade, was installed by 1958. It was rapidly followed by five more. The final closure, the Eastern Scheldt, took longer, delayed because the estuary was a spawning ground for a number of North Sea fish species and home to a prosperous oyster fishery, and those vested interests needed looking after first. The compromise was ingenious: a moveable storm-surge barrier that could be kept open in all but emergencies but that could be closed at need by shifting a series of massive gates the size of an office building. The whole thing was finished by 1986, and Queen Beatrix, in her speech christening the Delta Project, expressed a pious hope that no further defenses would be necessary.

The Greeks could have told her: hubris is hazardous, and often backfires. Less than a decade later, the country was reeling once again from flooding. This time it was a sneak attack, from the rear: in 1993 (and again in 1995) the flood waters came down the Rhine and the Maas (Meuse), a consequence of exceptionally heavy rains in the Swiss Alps and the Vosges and Ardenne mountains. More than a quarter of a million people had to be evacuated.

Exceptionally heavy rains caused it—but human malfeasance made it worse. By the 1990s the Rhine was no longer really a river, but an engineered waterway of levees, concrete embankments, locks, flow-control devices, hydro plants, weirs, and channels. Once the Rhine had meandered over an extensive flood plain; now, cut off from its natural controls, it flows twice as fast as before in a channel up to 26 feet deeper than before, dramatically increasing the power of its floods. In 1993 the Netherlands was forced to absorb the full effects of the Rhine's more powerful flow; neither the Germans nor the French, who exacerbated the damage, offered any compensation. The Dutch, polite to a fault, buttoned their lips and buckled down to work, devising what is now called the Major Rivers Delta Plan, increasing the height of certain dikes and allowing for the rivers to spill through others when certain emergency levels are reached. No one is tempting fate by saying the flooding is over. Instead, they are gloomily contemplating rising sea levels caused by global warming. And planning to build whole subdivisions of floating houses.

O f course, not all floods are major, and not all cause massive casualties. Floods are as familiar as summer. No place is immune, not even the great deserts—Sonni Ali Ber, one of the great tyrants of Saharan history, was killed not on the field of battle but in a flash flood after a freak rainstorm, a torrent that roared down a *wadi*, a gulley, and enveloped his entourage. Floods happen even in places inured to plenty of rain. For example, in July 2007 a month's worth of rain fell in a day in Yorkshire and the Severn River valley, followed by more torrential rains. Ancient cities like Oxford and Tewksbury were inundated. Drains were overwhelmed, rivers burst their banks, and at least a score of people drowned. Hundreds of thousands were made homeless, and, as *The Guardian* reported, many had to hire campers and trailers to park outside their ruined homes to deter looters. The cost was put at several billion pounds. Gordon Brown, the English prime minister, blamed the heavy rains on global warming; by coincidence, and by way of support, a study by a number of climatologists was issued the same week, asserting that the climate models had underestimated the amount of increased precipitation global warming would bring—rainfall would go up not 1 to 3 percent per degree as earlier suggested, the report said, but by a formidable 7 percent.[8] Flooding would become a fact of everyday life.

The most famous flood in English history, though, was long before the word "anthropogenic" saw its way into print. This was the disaster that struck the Bristol Channel in January 1607, causing the greatest loss of life from a natural catastrophe in western English history: somewhere between 500 and 2,000 people drowned when . . . something . . . surged up the channel from the open sea.

Whether this something was a tsunami or something else remains controversial. The tsunami theory is derived mostly from the publication of a tract shortly afterwards, which described the event in tsunami-like images, if in a disapproving, apocalyptic tone. The pamphlet, anonymously produced but clearly the product of a Puritan vision, was titled *God's Warning to His People of England*:

Many are the dombe warnings of Distruction, which the Almighty
God hath lately scourged this our Kingdome with; And many more
are the threatning Tokens, of his heavy wrath extended toward us. . . .
Then they might see & perceive afar off, as it were in the Element,
huge and mighty Hilles of water, tumbling one over another, in such
sort as if the greatest mountaines in the world, has over-whelmed the
lowe Valeys or Earthy grounds. Sometimes it so dazled the eyes of
many of the Spectators, that they immagined it had bin some fogge
or miste, comming with great swiftnes towards them: and with such a
smoke, as if Mountaynes were all on fire: and to the view of some, it
seemed as if Myliyons of thousandes of Arrowes had bin shot forth
at one time, which came in such swiftnes, as it was verily thought,
that the fowles of the ayre could scarcely fly so fast, such was the
threatning furyes thereof.

But there were no earthquakes in the vicinity, or volcanoes, and
there is little evidence of any major landslide. Nor is this in any sense
a subduction zone. To generate a 10- to 13-foot tsunami at a distance
of several hundred miles would require an earthquake of magnitude
7.5 or higher, which would surely have been felt somewhere along
the coast of England and Ireland.

More plausibly, the flood was caused by an extreme springtide
superimposed on a wind-driven storm surge focused up the Severn
Estuary. A wonderfully literate report written by Risk Management
Solutions, a California-based but British-owned consultancy, on the
four-hundredth anniversary of the flood had this to say:

The 1607 floods occurred just at the emergence of widespread liter-
acy in England, but before the scientific revolution of the mid-
17th century. The year 1607 marked the date of the first permanent
English settlement at Jamestown in America, and a number of ships
sailing to the New World were held up by stormy weather during the
month of January. Shakespeare was still writing and performing, and
the King James Bible began to be translated and edited. The language
of the King James Bible is exactly contemporary with the descriptions of the flood; the style in which the flooding is described is

vernacular, freeform and eloquent: *. . . let us fix our eyes upon theise late swellings of the outragious Waters, which of late now hapned in divers partes of the Realme, together with the overflowing of the Seas in divers and sundry places thereof: whole fruitful valleys, being now everwhelmed and drowned with these most unfortunate and unseasonable salt waters.*[9]

"Theise late swellings of the outragious waters . . ." Precisely. Jehovah couldn't have said it better himself.

But lest you think this all a historic curiosity, the same RMS report pointed out that a very similar storm and storm surge had recently (spring of 2007) been recorded in the Gironde estuary, in France; and that at the Balayais nuclear generating station, with its four reactors, the surge had reached 3 feet higher than had previously been thought to have been possible, and flooded several feet of the lower level of the gallery, forcing the authorities to progressively shut down all four reactors.

The French Nuclear Safety Authority admitted that there had been a Level 2 Emergency at the site. Although not confirmed, there were rumors that three out of the four primary cooling pumps were lost as a result of short circuits during the surge, and that the operators of the reactors had sent messages to the authorities warning of the potential for an impending catastrophe. On the Bristol Channel, the reactors at Hinkley Point and at Oldbury are also vulnerable to being flooded by extreme water levels higher than anticipated in the design of the facilities.

With lots more rain to come as the planet warms, that should concentrate their minds.

THIRTEEN

Vile Winds: Tropical Cyclones and Tornadoes

It reminded me of the semi-circle of marble columns around the plaza of St Peter's in Rome, but these columns extended all the way around, and the convection currents had pushed them 50,000 feet into the sky. It was exquisite.
—Aviator Max (Maximillian C.) Kozak
  in the eye of Hurricane Dot

B ecause of their short life, very small footprint, and utter ferocity, tornadoes cause damage that can seem capricious and sometimes deeply malicious. A tornado that touched down in
Johannesburg, South Africa, ripped the roof off a house and whirled
it two blocks away, leaving the occupants shaken but quite unharmed;
their neighbors' houses were untouched. In Barrie, Ontario, a tornado
sheared a house in two, peeling it open like a doll's house; in an
upstairs bedroom, an ironing board was still standing, the iron still on
it, as though ready for use. In Phoenix, Arizona, a tornado ripped up
a utility pole and hurled it through the windscreen of a pickup truck,
traffic light still attached; neither of the two occupants was injured, the
pole and shattered glass having been driven by inches from their faces.
In Utica, Illinois, a tornado killed eight residents who had taken shelter in a tavern; a nearby house was torn in two, but a bed of tulips at
the front door was still happily in bloom.

So often, it is the trivial things, or miraculous escapes, that catch the attention. Almost every twister leaves behind some curious anomaly. I myself have seen a wooden clothespin driven deep into a pine by the wind. A children's doll is driven feet-first into the trunk of a tree, but is otherwise undamaged; the whole roof of a house, still intact with its shingles and gables and gutters, more than 1,600 feet from the house it once adorned; a car containing two children hurled into the air and back to the ground, the children unharmed; a schoolhouse with eighty-five pupils demolished and the children blown 325 feet away, seriously frightened but unhurt; five railway coaches, each weighing seventy tons, moved 100 feet off the tracks, but still upright. . . . One house can be demolished, the one next door, a mere 3 feet away, untouched. Sometimes the funnel leaves the ground altogether, only to touch down again a hundred feet or so farther on—sparing one or two houses in a long row, with, always, apparently demonic unfairness. A tornado can rip asunder a house—or even a community—in seconds, and very little that is human-made can withstand its force.

A tornado is a vortex. The word is Spanish and is derived from the Latin for turn, *tornare*, which is what vortexes do. Tornadoes spin tightly, with awesome speed. This is the most violent of all winds. Just how violent is still unknown, because tornadoes routinely destroy even the most robust of measuring devices, even supposing one could be placed in a storm's unpredictable path. The highest wind speed ever actually measured within a tornado was near Red Rock, Oklahoma, in April 1991, when a twister was clocked at 286 miles an hour, much faster than a Category 5 hurricane like Camille, and it is possible that occasional tornado winds might even exceed 465 miles an hour. The energy within a single tornado is not much less than the 20-kiloton bomb dropped on Hiroshima.

The United States has the dubious honor of being far and away the most common locus for tornadoes, both in frequency and violence, with Australia an unenthusiastic runner-up. Other common enough spots are the Ganges basin of Bangladesh and the Yangtze

River valley of China. Tornadoes also happen on the great plains of Africa, occasionally in central Canada, and are not unknown in Europe—Aristotle described a tornado in the handbook he titled *Meteorologica*, and a tornado ruined parts of Rome in 1749. Occasional tornadoes can occur just about anywhere thunderstorms happen, which is pretty much every place on the planet; hurricanes can also spawn tornadoes. The tornado belt in the United States, known as Tornado Alley, runs in a swath across the Great Plains from north Texas through Oklahoma, Kansas, and Iowa to southern Minnesota. Somewhere between six hundred and a thousand tornadoes touch down in the United States every year. The Great Plains, as it turns out, are the perfect kitchen for cooking up tornadoes. This is because the eastern half of the continent is overlain in summer by warm, moist air coming in from the Gulf of Mexico, and the western states, where the prevailing winds are westerlies, are very dry—they are in the rain shadow of the Sierras and the Rockies. May is generally the worst month for tornadoes, but they can happen all summer.

America's worst tornado was in March 1925, when a twister roared at 60 miles an hour through a series of small mining towns from eastern Missouri to western Indiana, killing 739 people. But for sheer perverse capriciousness, the unfortunate victim has got to be a small town in Kansas named Codell, which was hit by tornadoes three years in a row on exactly the same date, May 20, in 1916, 1917, and 1918. The tornado-free May 20, 1919, must have been rather a big day in town.

Despite their ferocity, by hurricane standards tornadoes don't kill very many people. The worst tornado toll in recorded history was a Bangladeshi storm in 1977 that killed 900. Eleven of the 18 worst death tolls from tornadoes have been American. The highest in the United States was the 1925 Missouri twister, followed by another that killed 300 in 1974. And no, it's not true that tornadoes have an affinity for trailer parks.

T he strongest tornadoes are the creatures of mammoth and long-lived storms called supercells, whose winds are already rotating (they are themselves vortexes, albeit slow-moving ones), that may carry updrafts and downbursts exceeding hurricane strength, which is one very good reason pilots don't like them. Some of the other ingredients necessary for birthing tornadoes are warm, humid air near the ground, cold air at higher altitudes, and shearing winds. It is the humid air rising rapidly into colder air above that precipitates ice or rain, which in turn releases enormous latent energy, which then refuels the storm. Supercells almost always carry massive amounts of moisture, which often comes down as hail—many observers have reported what they call hail roar during a thunderstorm, the sound of billions of hailstones clattering together on the way to the ground.

Tornadoes can form very quickly and are very hard to predict. Warm air rising rapidly into colder air above is a necessary precondition, but if the warm air rises steadily and smoothly, tornadoes are actually unlikely. A much more likely result would be another series of rather weak thunderstorms. But if a shallow layer of just-warm-enough air hovers above the surface—warm enough to prevent the ground-level air rising—the potential is much greater for serious damage. Because if that cap is somehow moved or damaged, say by an incoming cold front, the pent-up warm air on the ground can burst through very rapidly. Then watch out. Tornadoes can appear in less than an hour.

Tornadoes are measured not so much by their velocity, which is so often unknown, but by their potential for mayhem. That is, because of their explosive but transitory nature, they are really only categorized after the fact, by the damage they have caused.

The commonly accepted tornado intensity scale is the Fujita Scale, first written in 1971 by Theodore Fujita of the University of Chicago, together with Allen Pearson, then director of the National Severe Storm Forecast Center. It makes ominous reading, even if Fujita was, in the upper reaches, somewhat stretching the limits of adjectival vocabulary; it ranges through "moderate" to "incredible." Fujita 1s (F1s) or "moderate" tornadoes range from 74 to 112 miles an hour, well up into the hurricane range, and will peel off roofs, overturn mobile homes, and push cars off roads. F5s are described as "incredible" and their winds range from 261 to 323 miles an hour, "with strong frame houses lifted off their moorings, cars flying about, trees debarked and even steel reinforced concrete severely damaged." And yet Fujita described one grade above incredible, which he called "inconceivable." These, if they ever occurred, would carry sustained winds of 324 to 378 miles an hour, but no one will ever know for sure, because all measuring devices would be destroyed, along with pretty well everything else in their path.[1] A quarter of all tornadoes are marked as "significant" (F2), and only 1 percent are F3s or above, the most violent categories.

The forward speed of tornadoes is usually around 28 miles an hour, but they can be nearly static or move much faster. Their paths are usually narrow, no more than several hundred meters and sometimes fewer. Length varies widely from not very much to dozens of miles. Most people's first sight of a tornado is its funnel shape, which always seems to be striking downward at the Earth, but this is an illusion. Tornadoes do form from the top down, but they aren't visible until they pick up debris from the ground—what you actually see is a grotesque mixture of soil, shrubs, fragments of trees, window glass, household effects, car bodies and sheets of plywood, metal roofing, flying barns, bits of houses, whole cows . . . and sometimes people.

A fully developed tropical cyclone is bigger, slower, and much longer-lasting than any tornado and can cause a swath of destruction that only a major earthquake can rival. Tornadoes are small and deadly; hurricanes (Atlantic cyclones) and typhoons (Pacific cyclones) are very large and very deadly.

Tropical cyclones can last for weeks; they are self-fueling storms, sucking up energy from warm sea surface temperatures. They can travel 5,000 miles and sometimes more; some of the bigger storms have been known to originate in Africa, coil their deadly way across the Atlantic to the Caribbean and the southern United States, turn north to traverse the western Atlantic, sometimes brushing by the Canadian Maritime provinces, and then catch the prevailing westerlies to pummel Europe, before disappearing into Siberia. Some cyclones are tightly coiled, but others can have a footprint as wide as Texas; Hurricane Ivan of 2004 was big enough to completely envelop the Gulf of Mexico before the eyewall made landfall in Alabama. The energy they carry is enormous—even a moderate hurricane releases enough energy in a single day to equal four hundred 20-megaton nuclear bombs; if converted to electricity it would be enough to power all of New England for a decade, with enough left over to run toasters all over the Canadian Maritimes. It can shift a billion tons of water before it runs its course; it is one of the engines that drive the ocean circulation.

And they're common. They happen every year. And they may be getting more intense as the Earth warms and the sea surface temperatures rise.

M ost of the casualties in hurricanes and typhoons are caused not by the wind itself but by flooding, either by storm surges or heavy rainfall; it is not surprising, therefore, that the highest death tolls are found in low-lying countries that are densely populated. Since records were kept, 13 typhoons have killed 10,000 or more people, and all of them occurred in Asia, 7 of them in Bangladesh (a 1970 typhoon killed 300,000 people, the worst toll in known history; another in 1991 killed another 139,000 people). The largest death toll from an Atlantic hurricane was Hurricane Mitch in 1998, which killed 9,200 people, closely followed by Fifi (1974, Honduras, 9,000 people killed). The storm that destroyed Galveston, Texas, in 1900 left 6,000 dead, and Georges, in 1998, killed 4,000. Katrina's death toll is still disputed, but at least 1,836 victims are known, with many still missing. By contrast, Hurricane Andrew of 1992, next to Camille the strongest storm ever to make landfall in America, killed only 62 people. The differences between the American and Asian tolls are due to lower populations, sophisticated warning and evacuation plans, and stronger buildings.[2]

Hurricane winds are not as strong as tornadoes, but if you're in the teeth of one you could easily be forgiven for thinking the differences are illusory. Hurricanes can strip the entire skin from a high-rise building, as one did in Charleston, South Carolina, leaving just the steel skeleton (most of the residents having got out, just in time for their furniture and belongings to be sucked into the void). Hurricanes can stir up the sea into ferocious battering waves 13 feet tall, coming on top of a three to six feet of storm surge and possibly on top of a high tide—the sea hurling itself at the shore 20 or more feet above normal, tossing boats around, demolishing wharves and jetties, destroying low-lying buildings; boulders as big as cars rolling around in a furious game of marbles. Hurricane Andrew stripped the antennae off the National Hurricane Center in Miami, forcing the forecasters to rebuild farther inland, this time in a concrete bunker half underground. A television reporter in Dade County, Florida, risking his life for his station, recounted seeing a 30-foot sign barreling down the center lane of an expressway and crashing into an overpass; he tried to get pictures, but his cameraman was blown away, along with his camera. An errant roof tile punched through a steel roof; a flying 2-by-4 penetrated a concrete block wall.

Hurricanes are categorized on a scale devised by Herbert Saffir, a building engineer, and Robert Simpson of the National Hurricane Center. The Saffir-Simpson Hurricane Scale is a 1-to-5 rating based on the hurricane's intensity at the time of sampling. Category 1 hurricanes range from 74 to 95 miles an hour; Category 5s, the most severe, from 155 miles an hour and up.[3]

It is a curious fact of wind that even in the very worst storms, and even in an open field with no obstructions, the velocity at ground level is effectively zero. When hurricane forecasters refer to surface wind speeds they really mean velocities at a standard height of 33 feet above the surface; from 33 feet wind is assumed to increase in speed with height, and generally does. The Saffir-Simpson scale, then, refers to sustained wind speed, measured over a full minute, at 33 feet. Ratings assigned to hurricanes by the weather service are used to give an estimate of the potential property damage and flooding expected along the coast from a hurricane landfall, but wind speed is always the determining factor, as storm surges depend to a considerable degree on the slope of the continental shelf in the landfall region, on the elevation of the nearby land, and on topographical peculiarities—is there, for instance, the possibility of a funnel effect, which would push a surge higher than normal?

Anything above a Category 2 is considered a major hurricane, likely to do considerable damage to buildings and the landscape. Only five Category 5 storms have ever hit continental America, two of them (at time of writing) in 2007 alone. The first was an unnamed storm that struck the Florida Keys in 1935, when the barometer fell to an extraordinary 892 millibars (26.35 inches). This Labor Day storm killed more than 400 people; some press reports claimed some of the victims were quite literally sandblasted, reduced to bones, leather belts, and shoes, but this is doubtful. The second was Hurricane Camille of 1969, which struck the Mississippi coast with sustained winds of over 190 miles an hour and a storm surge of 25 feet above the mean tide levels—a three-story wave rolled through Pass Christian, knocking over apartment buildings, and one appalled survivor who had retreated to his attic was forced to break the window and swim to a nearby transmission tower, from which he saw the water submerge the peak of his roof. He lived one and a quarter miles from the ocean. Camille is still the most intense storm ever known to have made landfall in North America; the winds were so strong— probably 200 miles an hour from Long Beach to the ironically named Waveland—that entire sections of the Mississippi coast just vanished. Although there had been plenty of warning, and evacuations had gone on apace, hundreds of people were killed and more than 14,000 homes completely destroyed. It probably didn't much encourage the survivors when President Richard Nixon ordered the dropping of 100,000 pounds of the pesticide mirex on the ravaged communities, in an effort to destroy the plague of rats that followed. Hurricane Andrew of 1992 was the third. Originally classified as a Category 4, it was Andrew that blew away the antennae and radar disks of the National Hurricane Center; the storm's rating was upgraded to a 5 more than a decade later, in late 2003. The most recent such storms were Hurricane Dean, in August 2007, which registered 906 millibars with winds of 184 miles an hour just before it struck the Yucatán, and Hurricane Felix, just a week later. But we're still counting. Felix will not be the last.

Katrina, the storm that mauled New Orleans in 2005, was briefly a Category 5, but was Category 3 when it made landfall. Ivan, of

2004, is still the only storm ever to have reached Category 5 not once, not twice, but three times, along its tumultuous course.

Not every part of the world is vulnerable to a hurricane or typhoon. Tropical cyclones always start, as should be obvious enough from the name, in the tropics. But they never start on the equator itself—there is no Coriolis force at the equator, and so no way to get a storm spinning. They always start in latitudes just high enough for the Coriolis force to be appreciable. But they never start at high latitudes either—the ocean is too cool there, the sea surface temperatures, or SSTs as the hurricane hunters call them, far too low. They may have their remote origins over land, especially where high heat and cool air produce thunder cells, but hurricanes proper never form on land—evaporated moisture is their fuel. Hurricanes hardly ever form in the southern hemisphere either—before they can properly organize they are broken up by the prevailing westerlies, which in the southern hemisphere are much closer to the equator, although in 2004, for the first time, Hurricane Catarina struck Brazil, the extraordinary product of high sea-surface temperatures, low vertical wind shear, and strong mid- to low-level blocking currents, leading to dire Brazilian forecasts about the coming apocalypse called global warming. Few tropical cyclones start in the smaller Indian Ocean either. Or if they do, they don't amount to much—the "fetch" of the ocean, the amount of sea available for a storm's nourishment as it travels, is too small. Still, in 2007 a damaging Indian Ocean cyclone hit the Arabian Gulf, threatening global oil supplies and an array of buildings that had been constructed with no thought to typhoon-scale wind forces.

Many of the Atlantic storms are born in the Sahara, as the superheated air of the desert meets the cooler air over the mountains, and then is energized as it drifts out into the Atlantic—and so the weather offices in Timbuktu and Niamey and in Abidjan, in the Côte d'Ivoire, and in Dakar, in Senegal, are the early warning systems for Atlantic hurricanes. Early summer Caribbean hurricanes, though still made up of African-born tropical waves, tend to form in the western Atlantic and in Caribbean waters, because the sea there is shallower and heats up faster. It is the late-season hurricanes that pass over the Cape Verde Islands before heading their relentless way westward.

In the eastern Pacific, most cyclones fizzle harmlessly in the colder water west of the Americas. Hawaii gets a few, but the prevailing easterlies steer most of them well south of the islands. The western Pacific is even more prone to cyclones than the Caribbean, and the "season" is year-long and therefore much more dangerous than the six months or so of the Atlantic. Pacific typhoons are born at sea and tend to be larger and better organized than their Atlantic cousins—the much greater stretch of the Pacific gives them substantially more room to mature than the smaller Atlantic. The classical Japanese description of a typhoon is *kamikaze*, or "divine wind"; Japan and the Philippines are frequently assaulted by three or more storms a year. Korea, China, and Vietnam are also vulnerable. To the south, Australia's northern littoral is under threat from typhoons in summer and fall, from December to about May; about a dozen cyclones a year form offshore, and some of them strike land. The worst storm in Australian history was Tracy, which struck Darwin on Christmas Day 1974.

One of the preconditions for the formation of a tropical cyclone is stable air, a local environment where the winds don't change very much. Hurricanes don't form in patches of turbulent or active air— powerful as they are, destructive as they are, they are also curiously vulnerable at birth. There must be very little wind shear, allowing the big cumulus thunderstorms to build vertically. Any strong wind above a hurricane will destroy it, either by tipping it over through shear, or by literally poking holes in the warm core tube, allowing the warm air to vent, which weakens it. Or they'll "plug the chimney" at the top, where the storm's vent is.

Another precondition is warm water. The ocean below must be at least 78.8 degrees Fahrenheit, but preferably 79.7 or higher. No real scientific consensus exists on why this number is the magic one. It has perhaps to do with the climatological factors governing tropical oceans, but which factors and precisely how are still unknowns. Temperatures can be higher than 78.8, but not lower—the higher they are, the greater the potential for damaging convection currents to occur. Higher temperatures don't increase the probability that a system will coalesce into a hurricane, but they will tend to make that hurricane more intense.

If these several preconditions are met, the winds of the passing thunderstorms, still just a tropical disturbance or tropical wave, will evaporate this warm water, and because of the Coriolis force the winds will lazily circle inward to the center. This causes a small vacuum, and the pressure drops, driving the warm, moist air upward. At about 2.4 miles above the surface, the vapor begins to condense into water, or into shards of ice and wet snow, and the act of condensation causes something deadly to happen: the heat and therefore the latent energy contained in the moist air is released. This energy is substantial—35 ounces of water will release enough energy to boil half a quart of water. This liberated energy then reheats the air, driving it still farther upward, creating even lower pressure below, and drawing still more warm, moist air into the atmosphere. Consequently, the pressure drops still further, more heat rises into the sky, and the system, now beginning to spin faster, becomes a self-sustaining storm, isolated from the air streams that surround it. If it remains coherent and well organized for at least 24 hours, that's when the hurricane centers of the world start to pay attention, because the precursor conditions for tropical cyclone formation have been met.

If the sustained winds reach 23 miles an hour, the system is declared a tropical depression. Then it is given a number. Tropical depressions are tracked, because they are embryo hurricanes.

These tropical depressions are already convection engines, their fuel provided by the warm sea water, which evaporates ever faster in the higher winds, causing ever lower pressures. After a few days, the tropical depression may escalate into a tropical storm. If the sustained winds reach 38 miles an hour, the meteorologists reach for their naming dictionaries and give the new storm a moniker. At 74 miles an hour, the threshold at which a tropical storm becomes a baby hurricane, pressure can drop very rapidly from the periphery to the center—pressure has been tracked to drop 38 millibars in 30 minutes in particularly severe storms.

Meanwhile, the whole system is moving, at speeds of up to 32 miles an hour. Atlantic hurricanes and Pacific typhoons both move generally westward at first, then curve northward, and eventually northeast. This is why, in the northern hemisphere, winds are strongest to the storm's right, where the directional speed of the storm's travel is added to its rotational speed. To its left, the forward speed mitigates the observed wind speed. The faster the storm is going the more exaggerated this effect. If a hurricane is going to hit, you should hope you're to its left.

When a small hurricane's convection pattern strengthens, the centripetal wind flow gains speed, the input of warm moist air continues to escalate, even more large-scale condensation occurs during the ascent, and enormous amounts of latent energy are released, which in turn result in stronger winds, which in turn lead to uplift of more warm moist air to condense and release ever more energy. . . . A mature hurricane never blows itself out, as long as there is warm water to sustain it. The evolution from depression to full-blown hurricane usually takes a good four days.

D espite all this data, the actual birth of a hurricane still remains invisible. What makes one system coalesce into a storm and another dissipate? A hundred Saharan thunderstorm systems drift into the Atlantic each year, but only a fraction ever become tropical disturbances. Of those that do, only another fraction become storms, and not all of those become hurricanes. Globally, tropical cyclones are still uncommon. In any given year, the number will vary from 40 to 110, with somewhere around 12 or 16 in the western Atlantic; of those, perhaps 4 or 5 will be defined as major. This may be changing: from 1951 to 2000, there was an average of 10 named storms in the Atlantic each year; for the past few years the number has increased to about 15, and may still be going up—there are more tropical depressions to start with, and the ocean is 4 to 6 degrees warmer than in earlier decades. In 2005, there were so many that the forecasters had to reset the alphabet—they had run out of names. But the moment a storm becomes a hurricane is still hard to see.

In theory, it should be easy to track the beginning of a hurricane: simply wind the film backwards. We already know how a mature storm (step C) results from its westerly track across the Atlantic (step B), which was caused by a tropical storm off the Sahara (step A). Why not then follow it backwards from C to B to A and then beyond to see how the whole thing began? With the vast array of data provided by the globe-encircling network of satellites, we seem to have plenty of information—those satellites can track phenomena to a resolution of a few meters. In practice, what you see when the film is unspooled is this: hurricane, smaller hurricane, tropical storm, tropical depression, thunderstorm, moist windy spot, then a set of weather conditions that in no way look any different from those that cause, well, nothing. . . . What is it that energizes some of these warm moist spots into hurricanes? No one knows. All they can say for sure is that it must be very small, because it is currently beyond our ability to detect.

The charts of historic hurricane tracks pinned to bulletin boards in hurricane centers everywhere need to be interpreted with great caution. Both the tracks and their numbers are truly unpredictable. Because hurricanes, like other natural forces, are chaotic systems, you cannot predict how many will happen by looking at what happened in the past. There may be apparent patterns, but they are illusions. So if you know that from 1951 to 2000 the average annual number of storms was ten named tropical storms, six of which became hurricanes, or if you know that last year there were fifteen named storms, and nine hurricanes, this average and this raw number are entirely useless to predict what will happen this year or next. After all, in 2005 there were more than thirty named storms, including the lamentable Katrina, and twenty-four hurricanes. In 2006, there were six.

Global warming may make things worse—higher sea surface temperatures make for more intense hurricanes. On the other hand, higher temperatures may also make for stronger high-altitude winds, and therefore more wind shear—and therefore fewer hurricanes. But it wouldn't do to count on it.[4]

FOURTEEN

Plague and Pandemic

And the first [angel] went, and poured out his vial upon the Earth; and
there fell a noisome and grievous sore upon the men. . . .
—Revelation 16:2

E bola, the deadly hemorrhagic fever, broke out in Gabon while I
was in neighboring Cameroon, not more than 124 miles or so
from Albert Schweitzer's celebrated leper hospital at Lambaréné. It
swept through two Gabonese villages with ferocious efficiency,
killing more than 80 percent of the people infected, before the news
of it got to the capital, after which it was swiftly isolated and suppressed by Gabonese and American medical teams. Still, everyone
was keeping a wary eye on the area, just in case. Ebola was definitely
not something you wanted in your take-home baggage; the notion
of this virulent little filovirus hitching a ride on an international
flight was enough to give the creeps to all the people at the World
Health Organization (WHO) and the American CDC (Centers for
Disease Control). I'd been especially cautious, since I had been heading that way anyway.

I'd been wanting to take one of the great excursions from the capital, Libreville, on the Transgabonnaise, the newly built, highly efficient, and ridiculously overcrowded cross-country railway, which ends up near Bongoville, which is not surprisingly, where M. Bongo, the more or less benign dictator of Gabon, was born. I was told that there was nothing very much to do in Bongoville but that the journey itself was worthwhile, passing as it did through some beautiful hills thickly covered with equatorial jungle. But I missed a connection, and in the end I drove from Ebolowa in Cameroon, over roads passable only with a four-wheel-drive vehicle, to the border village of Ambam. It is, after all, the same equatorial forest, with the same splendid panoply of exotic animals.

A Cameroonian gendarme who served a not exactly thrilling watch at the sleepy Cameroon–Gabon border post volunteered to take me on a hike into the bush, crossing as we did so the national frontier, which seemed to me sloppy of him. He promised to show me forest gorillas, but we saw none, though we did see their tracks and the plants they'd ruined in their foraging. The gendarme, a member of the Fang tribe named Jean-Bernard, was convinced that gorillas were savage creatures just waiting to rend humans limb from limb. Of course, his stories of their savagery might just have been told to puff up his own bravery, but he seemed convinced enough and started at every noise, of which there were plenty, and carried his carbine at the ready.

Naturally I was skeptical, having been reared on *National Geographic*'s misty sentimentality about the great apes and their gentle ways (and having seen the movie, in which the villains were always human poachers). This irked Jean-Bernard, who insisted I didn't know what I was talking about—typical white attitude, in his view. Not only do they kill humans, he said, but they ambush them, and drag their victims off into the forest.

"And they eat them?" I asked, even more skeptical.

"No, they eat plants," he said impatiently. "They just torment the people."

There were elephants everywhere, and we occasionally caught sight of them swaying through the jungle, as wrinkled and ridiculous as pantomime horses. A large python was furled around a tree limb in the sun, its head hanging down, peering at us beadily, for all the world as though it was about to say boo. A family of buffalo splashed through a creek. It was rather like a cartoon jungle, except that the animals really were wild and could bite or, in the case of the buffalo, gore.

I didn't know, then, that Ebola had helped push the gorilla into the "critically endangered" category of great mammals.

Green monkeys screeched everywhere. I was mindful that the green is thought to be the host animal for the Ebola virus, and though no one is sure, I was careful not to annoy them.

Jean-Bernard and I stood on a small hill, peering southward, where there would have been a view if the jungle hadn't been so dense. Not far away—an hour's drive on a good road, a day or two bashing through the bush here, an easy swing for a green monkey—had been the most recent Ebola outbreak. I asked Jean-Bernard about it.

He shrugged. "Not a problem in Cameroon," he said.

We were then standing in what he thought was Gabon.

"It doesn't spread?"

"No," he said. "You have to touch the dead to catch it."

He was a gendarme. "In your job, don't you occasionally have to touch the dead?" I asked.

He was horrified. "Never!" he said, and led the way back to the pathway.[1]

Two other nightmares, besides giving a filovirus a congenial environment in international air travel, keep researchers up at night. The first is that the virus might mutate out of stage 3 on the transmission scale to stage 4. In stage 3, an animal-to-human pathogen propagates for only a few cycles of human-to-human transmission before petering out, as Ebola does now. If the filoviruses made it to stage 4, as have, say, smallpox and syphilis, they would be viable for very long cycles of human-to-human transmission, which would make them incrementally more fearful. There is, after all, no known cure, and the mortality rate approaches 80 percent.

The second is that the virus might become more efficient at jumping from host to host. The worst scenario of all is that it mutates to become transmissible by aerosol—that is, on human breath. It is possible that a strain of Ebola, the Zaire virus strain Mayinga, may have already made this breakthrough. Just how Mayinga N'Seka, after whom the strain is named, contracted the disease remains uncertain, but some reports say she never touched anyone else affected and may have caught it simply by breathing.

The first recorded case of Ebola was in the southern Sudan, in the town of N'zara, on the fringes of the tropical rain forest. A worker in a local cotton factory fell ill, went home, collapsed, and died; in his death he transmitted the disease to dozens of friends and family. Within days it had rampaged through N'zara and made its way into the hospital in the neighboring town of Maridi. It was in that hospital that its true horror first became known. One of the nurses there who fled, and thus saved her life, described her terror as she watched people falling over and dying in agony, bleeding from every orifice, vomiting black blood as their intestinal systems disintegrated. The virus attacks almost every organ and tissue in the body; even the skin liquefies; the nurse described the victims weeping blood.

That the staff did flee, leaving the patients lying where they fell, might seem like a violation of medical ethics, but it was probably a good thing, for they had no other way of coping and had they stayed the number of cases would simply have gone on increasing. As it was, the epidemic ran its course and subsided without outside intervention, but not before killing 141 people.

Just a few months later an epidemic flared up in what was then Zaire (now the Democratic Republic of the Congo), in the region of Bumba, whose principal travel artery was the Ebola River. No one knows the index case, but the outbreak occurred in a hospital run by an order of nuns from Belgium. It was brought to the hospital by a schoolteacher who was not feeling well; within days it had spread to fifty villages, the spread made more efficient by the custom that bound family members to touch the dead before burial. One of the nuns who fell sick was taken to Kinshasa's University Hospital, where she was cared for by Nurse Mayinga, using the usual precautions against infection. It didn't help—somehow the virus jumped into her body and started its destruction. It took several days for Mayinga herself to be admitted to a hospital. She died two days later, but not before WHO had been alerted and the Zairian dictator, Mobutu, had the army seal off the hospital. No one who contacted her contracted the disease, and after killing 318 people with an 88 percent mortality rate, the virus did as the Sudanese one had done. It simply petered out.

So far, only two members of the virus family Filoviridae have been identified: Ebola and Marburg. Four substrains of Ebola have been found: Ebola Zaire Mayinga, Ebola Sudan, Ebola Reston, and Ebola Tai. The first two have already been described. Reston, alarmingly, was found in a research lab in Reston, Virginia, a heartbeat from Washington, D.C. in 1989 and is so far the only strain not to cause serious disease in humans, though it is fatal to monkeys. It may also be aerosol-infectious. Ebola Tai is from the Ivory Coast. Its virulence is unknown, since there was only one case, a Swiss researcher who contracted the disease in 1995 while working with chimpanzees, and she recovered.

The Marburg virus is named after Marburg, Germany, the site of the first outbreak, where workers from Germany and Yugoslavia, who were handling tissue from green monkeys, sickened and some died. Again, after the initial outbreak, the virus fizzled, though it was subsequently identified in South Africa when a traveler coming from Zimbabwe became ill and passed the virus to a number of his companions and at least one nurse in Johannesburg. No one died that time. Then, in August 2007, Marburg broke out again, this time in Uganda. The victims were miners who had re-opened an illegal lead and gold mine inside the Kitomi Forest Reserve. The suspected vector this time was bats. About five million of the creatures live in and around the mine; their fecundity alarmed WHO, which issued a bulletin calling the outbreak a global threat to public health. This may seem a little over the top, since there were only two cases, but the severity of the disease and the high mortality it caused set off alarm bells.

Filoviruses aren't by any means the only pathogens to have made the jump from animals to humans. In the last 30 years, 38 "new" illnesses have passed over to humans from a bewildering array of wildlife. Mark Woolhouse, of the University of Edinburgh, has been tracking the phenomenon, and has listed 1,407 pathogens—viruses, bacteria, protozoa, and fungi—that can infect humans. Of those, 58 percent come from animals and 177 are considered to be emerging, or possibly re-emerging. About 1 new infectious disease emerges every year. Most, Woolhouse says reassuringly, will never cause pandemics.[2] But some surely will.

One that already has is a retrovirus subtype, one of the "lentiviruses" (slow viruses) called the Human Immunodeficiency Virus, or HIV, the cause of the AIDS pandemic. Wild theories exist about the origins of HIV—including that it was spread by a promiscuous flight attendant (a gross libel on this hapless Air Canada employee, since the study on which the identification was based was later withdrawn by its author), that it was caused by a polio vaccination campaign gone bad, or that it was a conspiracy against African men by nefarious and unnamed racists. This last canard surfaced as recently as October 2007, put about by no less a personage than the Catholic Archbishop of Mozambique, who suggested that condoms laced with HIV were being given to Africans to reduce their numbers. The truth is rather more prosaic. Retroviruses have been found in a number of animal species, including horses, sheep, and cats. But the real vector was probably the same animal that hosts Ebola, the green monkey (also sometimes known as the sooty mangabey). An SIV (or Simian Immunodeficiency Virus) that is very close to HIV has already been found in green monkeys in Cameroon, and it is known that chimpanzees have been infected by greens, so the virus can cross at least one species barrier. It was probably transmitted to humans via "bush meat," as human hunters killed and ate infected monkeys. (Such transfers are still happening: *The Lancet* reported in 2004[3] that a survey of 1,099 men in Cameroon found ten infected with SFV [Simian Foamy Virus], an illness similar to SIV.)

The first identified victim of HIV was an adult male living in the Congo, near Brazzaville. The blood sample showing the virus was taken in 1959, and a lab analysis in 1999 suggests that HIV was introduced into humans around the 1940s or even earlier, possibly as far back as the end of the nineteenth century. (The year 1930 is the date suggested by a Los Alamos National Laboratory study in 2000 by

Dr. Bette Korber.) The earliest case in the developed world was a St. Louis teenager, who died in 1969.

In the end, it doesn't really matter how it got out of Africa (after all, *ex Africa semper aliquid novi*—there is always something new coming out of Africa), or even if it did. What matters is the grim reality.

HIV's very slowness conspired against attempts to stop it. By the time the first cases were identified and named, in populations of gay men in New York and San Francisco, the disease that it caused was already beginning its inexorable march across Africa. That it was at first written off as a "gay disease" and a "lifestyle disease" further doomed realistic attempts to track its course in a majority population that considered itself immune.

The current pandemic probably started sometime in the 1970s. By 1980 it was in every continent but Asia, and a few years later it was spreading there too. By the time anyone knew what it was, or began to talk about it, it had probably affected some three hundred thousand people, and was spreading rapidly.

Its penetration in Africa, where it was from the beginning a heterosexual disease, was most severe. The vector was the long-distance trucking routes.

Not long ago I went to see this vector for myself. Just outside the crumbling Zambian city of Livingstone I found a truck stop. The signs said, "Zambezi Rest Stop, Café, Good Food to Eat." The forecourt was pitted and stained with oil. The café itself was crumbling concrete brick, tropical weeds spilling from the eaves. About thirty huge trucks were parked haphazardly in the lot, some of them with motors still rumbling; others, with air conditioning that should have been rumbling, were quiet. Outside, the usual impromptu market, a few piles of tomatoes, bags of oranges, and stacks of Made-in-Kenya candies for sale. Knots of men stood around, shouting at each other, mostly in Swahili, the lingua franca of the Trans-Africa Highway. This is what Cecil Rhodes's great megalomaniacal dream of British Red from Cape to Cairo had come to: a thin ribbon of sometimes paved roadway peopled by the gypsies of our time, long-distance truckers. This is the only way to get by road from one end of Africa to another, from Cape Town to Cairo (almost—the road is still

interrupted in Sudan, and in any case most of the trucks go no farther than Nairobi).

This is the AIDS Highway.

The Zambezi Rest Stop Café is next to the Zambian customs extor-
tion shed, where functionaries delay trucks, sometimes for days, while
they search for something useful to steal. The truckers, of course, are
well aware of this but pay it little attention. Everyone understands that
a little goes here and there for squeeze, for "duties" duly stamped and
restamped. And they're in no hurry. Another day makes no difference.
They spend the time in Livingstone's bars and cheap hotels, and with
Livingstone's profusion of AIDS-ridden prostitutes.

In the towns and villages across Kenya, Tanzania, Uganda, Malawi,
Zambia, Zimbabwe, and South Africa, the virus inexorably spread. The
disease made its way south to the Limpopo along this highway, spilling
down into South Africa and into the teeming townships of Africa's
most muscular city, Johannesburg. And this same highway is how it
made its way north, to the Muslim countries.

Thousands are dying in the villages, where AIDS is called "slims
disease"—because that's what it does, emaciates. In some places a
whole generation of young children are born infected, their parents
dead or dying; travelers report villages in the countryside where no
adults are left alive. "This is a deadly thing," a public health nurse told
me in Tanzania, pushing a sheet of mortality statistics toward me. "The
deadliest thing, Africa's bane." How is it to be stopped? No one in
Africa has the resources. Africa has so very many deadly scourges.

Late in 1996 there was an optimistic announcement: the incidence
of new cases in Uganda, one of the worst hit, was dropping. But then
the statisticians got busy, and the optimism faded: the drop was attributed to the fact that most people who could get the disease already
had it.

AIDS also spread down the secondary trucking routes to Dar es
Salaam and Mombasa, and thence along the coast. Mombasa is the
port of entry for goods for Uganda, Rwanda, Burundi, and eastern
Zaire. In Busia, on the Kenya-Uganda border, every trucker and
hooker tested in a random survey was HIV-positive. Sixty percent of
the town was infected. By 1991, seven million people in Africa were

infected, a third of them with full-blown AIDS. A decade later, that number had more than doubled. Throughout Kenya, in rural areas, one in eighteen is infected. In cities, the figure rises to one in nine. In some areas, Kenya's own statistics suggest that up to a third of adults are HIV-positive.

After I left Livingstone, I checked in with the truckers wherever my route intersected the highway—Mbeya, Morogoro, Dodoma, Arusha, Nairobi. The routiers, it seemed, had two conflicting views of AIDS, when they had a view at all. It either does not really exist—people die of all kinds of things in Africa—or, alternatively, AIDS is a way foreigners have of putting another one over on Africans. Curiously, the villains of choice here seem to be the French, who are widely thought to have started it all.

None of the truckers seem to believe that you can be killed by doing what comes so naturally, so the prospects for prophylaxis are dim. There is no sex education to speak of. Indeed, in the villages there is hardly any medicine at all, only the occasional visiting immunization clinic. Making more children is a God-given right, and so are other women when one's wife is not available. As so often in Africa, the burden of change must fall on the women.

For too long, African governments denied the problem, sensitive to the implied outsider slur of sexual promiscuity. Some of them still do. South Africa's president, Thabo Mbeki, was still in 2007 an HIV denier, his AIDS policies—or lack of them—amounting to criminal negligence.

The headlines continued to get more dire. Stephen Lewis, the UN's special envoy for AIDS, went hoarse trying to persuade authorities everywhere that this was a global emergency. Everyone agreed, but little was done. If 20 million had died in the previous 20 or so years, another 65 million were forecast to die before 2020, more than all the people killed in all the twentieth century's wars. AIDS is now the world's fourth-leading cause of death. Some authorities—the NGO Global Health, for example—say the toll will reach 85 million by 2025. Others put the figure at 100 million. The peaceful little African country of Botswana had in 2007 the dubious distinction of being the country worst hit by AIDS: 39 percent of the population were infected.

Nathan Wolfe, of the University of California Los Angeles, together with several colleagues, is the researcher who proposed the five stages in the evolution of diseases from purely animal to purely human (of which Ebola and Marburg are stage 3, as reported, and AIDS now stage 5). His results were published in the journal *Nature* in 2007.[4] He found that most of the major infectious diseases, including some now not found in animals, are relatively new in human history, emerging only after the development of agriculture. The reasons were probably that farming made increased human densities possible, and of course farmers domesticated more and more formerly wild animal species.

The majority of these diseases, he found, came from strains that had been found only in Old World animals, transmitted to humans from warm-blooded vertebrates, mostly mammals, though flu and falciparum malaria come from birds. Hardly any were contracted from animals in the "New World" of the Americas, mostly because far more animals were domesticated in the Old World than the New. Only Chagas' disease (and possibly tuberculosis and syphilis) began in the New World. All of which, of course, greatly facilitated the European conquest of America.

Counterintuitively, most infectious diseases are not tropical in their origins. A somewhat higher proportion of stage 5 diseases are temperate zone ones, not tropical. Also, more temperate than tropical diseases are acute rather than slow or chronic—that is, the patient dies or recovers within weeks—whereas tropical diseases tend to be slower and less acute. None of the tropical diseases are "crowd epidemic diseases," capable of existing only in regionally dense populations, whereas most of the temperate diseases fit the definition. Finally, domestic animals were the source for eight of the fifteen temperate diseases studied (diphtheria, influenza A, measles, mumps, pertussis, rotavirus, smallpox, and tuberculosis). The others came from apes (hepatitis B) or rodents (plague, typhus) or from unknown sources (rubella, syphilis, although syphilis, once known as the Great Pox or as "lues," after its most famous victims, the many French kings named Louis, probably mutated from yaws, a generally non-fatal tropical skin disease).

Stage 1 in Wolfe's scale are microbes present in animals but not detectable in humans—most animal pathogens never make the crossing.

Stage 2 are pathogens that have been transferred from animals to humans but that have not made the next step of transferring themselves from human to human. Rabies is perhaps the most familiar example, but West Nile encephalitis, transmitted from mosquitoes, which in turn get it from infected birds, is similar. Pretty obviously, the more infected animals there are, the higher the likelihood of humans contracting the disease. (In North America, West Nile has already reduced the native crow population by 45 percent; five other bird species are facing similar declines.)

Stage 3, which is where Ebola and Marburg can be found, as well as, say, anthrax, are animal pathogens that can be transmitted in human-to-human cycles, but that last only a few of those cycles before attenuating. Such pathogens are typically virulent and greatly to be feared, for if they made the transition to the next two stages their effects would be devastating. Reassuringly, barriers to evolution do exist. Some of them are obvious enough, such as the differences between animal and human behavior (as Wolfe puts it, deadpan, humans seldom bite live animals). Others include obstacles within the human tissue system and the pathogen's need to mutate further before acquiring more mobility.

Stage 4 are pathogens that move from animal to human and can then undergo long cycles of human-to-human infection. Wolfe has identified three sub-stages here: pathogens whose spread is still mostly direct animal to human (yellow fever is an example); pathogens whose spread is equally divided between animal to human and human to human (dengue fever is a new and dismaying example of a hemorrhagic fever); and finally pathogens whose greatest spread has become human to human (there are many examples, such as influenza, typhus, and cholera).

Stage 5 are pathogens whose spread is now exclusively human to human, either through historical inheritance from other species (often from chimps and monkeys, which then evolved out of the infectious range) or through mutation—smallpox, measles, and syphilis being gruesome examples, and AIDS another. Smallpox, now eradicated, was particularly lethal, with a mortality rate of up to 40 percent, spread by contact with skin lesions or via the respiratory tract. Some reports indicate it may have killed better than 50 million people, and maybe as many as 100 million, in its long human history; the virus is very stable and mutates very little. It was smallpox that, indirectly, led to the development of vaccines: the word comes from the word "*vacca*," Latin for cow, because it was discovered that milk-maids who came down with cowpox, a milder version, never developed smallpox, which led a medical experimenter named Edward Jenner to come up with a radical notion (though one that would get him into serious trouble with human rights authorities in modern times): he deliberately infected an eight-year-old boy, not his own son, with cowpox to see if he could subsequently give him smallpox. Fortunately for the boy, John Phipps, it worked.

Novel stage 5 organisms are actually being encouraged by modern industrial civilization—international travel makes their spread easier, and so do blood transfusions; the existence of the bush meat trade; the agro-industry (which by traducing natural grazing and feeding habits has given us, among other nastinesses, mad cow disease); the production and mere existence of vaccines; the presence in the population of susceptible pools of elderly, antibiotic-treated, and immuno-suppressed individuals; and, of course, deliberate genetic tinkering and bio-warfare labs.

New diseases are always emerging, and old ones re-emerging. Tuberculosis, once thought to be on the way to extinction, has re-emerged in a more virulent, and occasionally not-treatable, form. It had existed for millennia—the ancient Greeks called it pthisis, a word that described a living body shriveling on a flame. The Romans called it consumption, a name that persisted into modern times. It is also known as the White Plague and has been somewhat romanticized by writers—Thomas Mann wrote about it, Keats and the Brontë sisters died of it, so did Molière, Voltaire, and Chekhov; Verdi's heroines always seemed to expire from it in more or less picturesque ways. But in the 1980s it began a resurgence that coincided with the AIDS pandemic (AIDS makes a patient more likely to die from TB). In

2006, WHO declared drug-resistant tuberculosis (XDR-TB) to be a global emergency, made worse by international air travel, for the disease spreads through coughing and sneezing, but also through yelling and even laughing. Millions of people are now thought to be infected with "drug-sensitive tuberculosis," which could easily mutate into the more deadly drug-resistant form.[5] Then in November 2007, the Washington-based Forum for Collaborative HIV Research issued a press release in which it made the following comments: "The largely unnoticed collision of the global epidemics of HIV and TB has exploded to create a deadly co-epidemic that is rapidly spreading in sub-Saharan Africa. . . . Approximately one third of the world's 40 million people with HIV/AIDS are co-infected with TB. . . . The mortality rate for HIV-TB co-infection is five-fold higher than that for tuberculosis alone. . . ."

> Ring a ring o' rosies
> Pocketful o' posies
> A tishoo! A tishoo!
> All fall down!
> —Children's nursery rhyme

Never mind that the above children's chant, widely thought to be a coded reference to the bubonic plague ("rosies" were the plague's lesions, "posies" the flowers carried to mask the stench, "tishoo" the characteristic sneezing of plague victims, "all fall down" obvious enough), is almost certainly just a skipping rhyme and nothing more. The fact that it is so widely believed to refer to plague is a sign of the terrible fear preserved in folk memory for what happened in antiquity—plague was, quite literally, the death that changed the world.

The first reference to plague in world literature is in the biblical book of Samuel (1 Samuel 5:5–6:18), which mostly contains a good deal of fretful to-ing and fro-ing about the Ark being hauled off by the Philistines, but contains two key sentences: "[T]he hand of the lord was against the city with very great destruction; and he smote the men of the city, both small and great, and they had emerods in their secret parts. . . ." And this passage: "Then they said, What shall be the trespass offering which we shall return to him? They answered, Five golden emerods, and five golden mice, according to the number of the Lords of the Philistines: for one plague was on you all, and on all your lords."

It takes a bit of imagination to see in these passages the ravages of bubonic plague—except for those "emerods," variously rendered in modern versions of the Bible as hemorrhoids, growths, swellings, and other excrescences. One of the characteristics of bubonic plague, and the one from which it derives its name, is the presence in its victims of agonizingly painful egg-sized swellings in the lymph nodes, especially the groin and the armpit, that are called buboes. It is unclear why the Israelites were advised to make five golden hemorrhoids, or why those should placate the Lord, who had inflicted them in the first place, but there you are.

The next time these buboes show up in historical writing is from Arabic sources describing the ravages of plague in the Byzantine empire of Justinian (482–565), followed shortly thereafter by apocalyptic references written in Syriac by John of Ephesus. After that there was a flood of lamentations, from Procopius, Agathias, Gregory of Tours, Paul the Deacon, and Bede, among others, all describing, in sometimes haunting images, the cataclysm that was facing the ancient world.

That cataclysm changed the world for good, if such a phrase can be used for such an evil demon. For one thing, plague weakened the remnant Roman empire's resistance to the armies of Muhammad, as well as to the predations of the Huns, Goths, and Visigoths. It was the subsequent collapse of the empire that birthed the nations-to-be of Europe and, by weakening resistance to Islam, allowed the rise of the Islamic superstate that dominated so much of the Middle East, Byzantium, Africa, and Spain for so many centuries. It also led to a great sea change in Christian thinking, which until then had emphasized humbleness and submission to an omnipotent but generous god, but was thereafter dominated by the notions of guilt, sin, and a much more wrathful Lord (Islam, for its part, interpreted the plague as God's way of punishing infidels). On the upside, Justinian's plague was probably the death knell for European slavery—so many laborers were killed that a premium was placed on any surviving skill, and workers could demand what conditions they liked.

Justinian himself was one of the great figures of antiquity, and one of the favorite "what ifs" of historians is to speculate what he would have accomplished had not an innocuous little bacterium evolved into something far more deadly. Most of his biographers make a good deal of his humble origins—he began life as a peasant in a remote Balkan hill town—"although the astonishing social mobility of the times is perhaps even better exemplified by his wife Theodora, a bearkeeper's daughter and prostitute turned racy comedienne, who once lamented the fact that Nature had constructed her so that she could only have sex via three orifices."[6]

Humble or not, Justinian had large ambitions. In 532 he set out to reconstitute the Old Roman Empire, much of which had been variously occupied by Franks, Goths, Vandals, Visigoths, Celts, and more, many of them Roman-trained in military affairs but entirely lacking any Roman notion of good governance. He accomplished much in a short time. He was able to recapture much of Italy. He established a new capital at Constantinople, later Istanbul. He built the magnificent Hagia Sophia, now of course a mosque, and put together the *Corpus Civilis*, the founding document of western legal theory. His quest to build a new empire seemed assured. Alas, *Yersinia pseudotuberculosis*, a bacterium that caused a mild illness resembling flu and that was transmitted through food and water, had come across a superior method of getting itself about: the flea. Inside the flea, something deadly happened to its new passenger. In its efforts to defeat the flea's natural defenses, the bacterium acquired a new gene, PDL, that allowed it to survive in the flea's gut. This turned it from *Yersinia pseudotuberculosis* into a new entity called *Yersinia pestis*, or simply *Y. pestis*. Ten years after Justinian started his quest to reconstitute the empire, in 542, rats brought fleas, and fleas brought *Y. pestis* and its plague, to his capital.

Bacteriologists at the U.S. National Institute of Health's Rocky Mountain Laboratories in Montana have isolated the PDL gene and have sequenced the whole genome of *Y. pestis* itself. They have speculated that sometime between 1,500 and 20,000 years ago the bacterium picked up the gene from other bacteria found in soils by a process called horizontal transfer, "somewhat akin to bacterial sex." "Our research illustrates how a single genetic change can profoundly affect the evolution of disease," as one of the bacteriologists, Joseph Hinnebusch, put it.[7]

The bite by an infected flea leads to the insertion of literally thousands of bacteria into a victim's skin. They migrate through shallow skin lymphatics to regional lymph nodes, where they rapidly multiply, causing widespread necrosis, massive and painful swellings, and septicemia that can quickly lead to shock, vein blockages, and coma, followed quickly by death. In heavily populated areas the plague mutated into pneumonic and septicemic plague, meaning that it could be spread by human contact and even by human breath, greatly exaggerating and facilitating its spread.

The mutation that caused Justinian's plague probably started in East Africa, perhaps in the highlands of Ethiopia. The rodents that carried it reached the harbor town of Pelusium in 540. From there, it quickly spread to Alexandria and thence, by ship, to the capital and the rest of the empire. Justinian himself survived, but as much as a third of the population died, maybe 20 to 25 million people, though some put it even higher, the favorite remedies of the time—amulets, saints' relics, butter rubbed on the buboes—having lamentably failed to work. All trade ceased, and civic life collapsed. The Dark Ages ensued.[8]

The human species is nothing if not resilient, and fecund. As the centuries passed, the population recovered, then exceeded what it had been, and continued to grow. The British Isles, having suffered first the anarchy that followed the departure of the Romans, and then the plague, built its churches and monasteries and slowly climbed back to some sort of civil order. Rome lost its civil authority and its political power, but the medieval popes, especially Innocent III, more than filled the political vacuum. The great cathedrals of Europe were erected to God's greater glory—the Ste. Chapelle in Paris and Notre Dame, Chartres, the great abbeys of England. Islam was triumphant in the east and overran all of Saharan Africa, but its wave of conquest washed up against Charles Martel at Poitou, and they were pushed back over the Pyrenees, leaving us with the *Song of Roland* and the legends of the magical bones of St. James of Compostella. Charlemagne took the crown from the Pope and the Frankish empire was born, while in Spain the former Almoravid Berbers constructed some of the most sublime architectural masterpieces in human history. Russia was ruled by the Golden Hordes, and their rule was not as terrible as the later propagandists of Ivan, himself the terrible, would later assert, and Muscovy prospered under the Tartars. In Africa, the golden empires of the sub-Sahara, Mali the Great and Golden Ghana itself, were beginning their climb to eminence, and their gold was beginning to make its way across the desert to the counting houses of the Maghreb; scholars made their way to the eminent schools of Fez, Timbuktu, Alexandria, Mecca; Islamic savants laid the foundation for modern medicine and mathematics. Far to the east, in Cathay, the sons of heaven went about their business of ruling their people, uncaring about the world beyond their borders.

The plague bacillus, however, was only resting. Still lurking in the rodents of East Africa, it had found another reservoir in the Kingdom of Heaven itself, in south central Hunan province, later the birthplace of the tyrant Mao, now one of the most industrialized parts of modern China. Somewhere around 1330 plague took root among the nomads who lived near the border of Guizhou, and from there it began its terrible travels up the old trade routes. By 1345 it was in the lower Volga regions of the Caspian Sea and took less than a year to make its way through to the Black Sea, the gateway to Europe. Plague epidemics broke out in the Caucasus and the Crimea that year. Another year found it in Constantinople and in Alexandria, across the sea; by the winter it had reached Cyprus, Sicily, and the Italian mainland.

We know how it got there from the Black Sea, through human enterprise and human folly. In the summer of 1347 Genoese merchants were doing business and then relaxing in the lush little resort town of Kaffa, on the Crimean peninsula, when the city was attacked by a band of Tartars from the Khanate of the Middle Volga. They had brought plague with them, and the siege failed when most of the attackers perished, but in a fit of rage, before the remnants departed, they hurled the diseased corpses of their comrades over the city walls. The Genoese fled as soon as they could, taking ship in twelve separate vessels bound for Sicily and Italy. By October, three vessels

had reached the Sicilian port of Messina, most of the passengers dead
and the remainder dying, "sickness clinging to their very bones," as a
contemporary report put it. They were forbidden to come ashore, but
no one thought to prohibit the rats, which scuttled into the city with
their deadly cargo of fleas. Within two months, half of Messina was
dead. Rome and Bologna and Florence followed in months. A third
of all Florentines died in the first six months. Marseilles saw its first
cases in January 1348, and it had reached Paris by the spring,
Germany and the Netherlands later the same year, London by
December. Norway saw it in the spring of 1349, and from there it
made its way to eastern Europe (1350), mysteriously skipping Poland,
and finally back to Russia in 1351. So many died in France that the
Pope, in desperation, declared the entire Rhone River to be "conse-
crated ground" so the dead could be hurled into it with some hope
of salvation. Within two years half of all Britons were dead, the toll in
continental Europe reaching and passing 15 million, with a final toll
that was probably 20 or 30 million, about a third of, well, everyone. A
Vatican report in 1351 put the toll at 24 million; it is probable that
another 20 million died before the plague ran its course by the end
of the century. No one knows how many people died in the Middle
East and Asia. More than 100 species of animals were affected too,
with similar mortality rates.

No one knew why any of this was happening. No one could yet
see the "animalcules," as the Dutch scientist Antonie van
Leeuwenhoek called the microorganisms he was the first to discover
with his brand-new microscope. No one even knew there could be
such things. Agriculture collapsed, and famines made everything
worse. Government collapsed along with farming, and so did every
industry but the church. Boccaccio, in the *Decameron*, bore witness to
the civic disintegration that followed. "Kinsmen held aloof, brother
was forsake by brother, oftimes husband by wife, nay, what is more and
scarcely to be believed, fathers and mothers were found to abandon
their own children to their fate, untended, unvisited, as though they
were strangers. . . ." Processions of flagellants shuffled through the
silent streets, flogging their own backs with whips, blood streaming to
their feet as they shrieked and cried, to the despair of the people who

watched them pass. Some of those blamed the Jews, always the target of choice, and pogroms followed, many Jews fleeing to the safe haven of Poland, where they were welcomed by the Polish king. In a vain attempt to restore order and to provide a rational cause for the insanity, a papal commission declared that the plague was not the work of an angry god, or the Jews, but that the Earth had passed through a hot, virulent miasma in the heavens, due to the unseemly conjunction of Mars, Jupiter, and Saturn in the House of Aquarius; the comforting thing was that the conjunction had passed and all would soon be well.

Folk remedies were as ineffectual as the Pope's reassuring edict. Bathing in urine was a favored specific; so was the practice of placing a dead animal in the house to attract the plague away from its human occupants. Drinking pus from suppurating sores was another popular device, as was the application of a live toad to the buboes. When all these failed, flagellation seemed the only thing left to try. But since the flagellants were forbidden to wash, that didn't do much good either.

Fleas, of course, were all too common. Many people, even the relatively well to do, cohabited with domesticated animals in a state of hygiene that would horrify a modern mother. Two centuries before the plague, a description of the burial of the martyr Thomas à Becket clearly showed the state of personal cleanliness. As his body was readied for burial, he was undressed. First they took off his outer garment, a capacious brown mantle. Underneath were found a white surplice, then a coat of lamb's wool, a woolen pelisse, another woolen pelisse, the black robe of the Benedictine order, and finally, next to the skin, a short, tight-fitting suit of coarse haircloth, covered on the outside with linen. When they took that off, and it was exposed to the chilly English air, so disturbed were the myriad creatures living in the hair suit "that it boiled over, like water in a simmering cauldron."[9]

Not everyone agrees that the Black Death, as this episode was named, was bubonic plague, or at least not only bubonic plague, or that the rat and its fleas were the real vector. They might not agree with astrophysicist Fred Hoyle's rather out-of-left-field notion that the Black Death actually originated in outer space (perhaps he had read the Pope's contemporaneous report from the 1340s), but Hoyle, and his collaborator, Chandra Wickramasinghe, did query one rather awkward fact in a 1979 book: how did the damn rats move so fast? "To argue that stricken rats set out on a safari that took them in six months not merely from southern to northern France, but even across the Alpine massif, borders on the ridiculous. What remarkable rats they were! To have crossed the sea and to have marched into remote English villages and yet to have effectively bypassed the cities of Milan, Liege and Nuremberg. . . ." Indeed, as Norman Cantor put it in his book on the plague, "it is this provocative picture of these rodents scurrying inland from port cities and making long journeys through the countryside at great speeds, [such] that most of western Europe was in pandemic conditions within a year of initial contact, that raises skepticism of the Black Death's exclusive identification with bubonic plague."[10]

The British historians Susan Scott and Christopher Duncan, in their *Biology of Plagues*, also argued that death spread through Europe much too rapidly to be caused by *Y. pestis*. Their nominee? An Ebola-like virus. James Wood, an anthropologist from Pennsylvania State University, used statistical methods to make the same point: "An analysis of monthly mortality rates [among priests] during the epidemic shows a 45-fold greater risk of death than during normal times, far higher than usually associated with bubonic plague."

Cantor himself leans to the theory that since some victims died without fever and without buboes, the Black Death "involved, or was even exclusively, a rare and virulent form of cattle disease, or anthrax," though he does allow a role for both.

More recent results have proved that whatever else may have been involved, bubonic plague was certainly a major culprit. In 1998, molecular biologists Michel Drancourt and Olivier Dutour of the University of the Mediterranean in Marseilles identified *Y. pestis* DNA in exhumed human teeth dating from 1590 and 1722, which were other plague years. Two years later, they reported a similar finding in remains dating from 1348.[11] That seems to be that.

Plague didn't just disappear. A third pandemic began in China in the nineteenth century and killed more than a million people in a dozen countries, including North America, before it petered out—though not before leaving a reservoir of the disease active in animals (gophers, rats, squirrels, and other rodents) in the American southwest, where it remains, from the Pacific coast to the great plains, and from southwestern Canada to Mexico. Another outbreak was in Manchuria in 1910–11, infecting some 60,000 people with a depressing 100 percent mortality. A small outbreak occurred in Los Angeles in 1924–25, but didn't spread far. Ever since, a dozen people a year are infected, mostly in rural areas; globally, perhaps 1,000 to 2,000 cases occur every year. The last real outbreak was in Surat, India, in 1994, in a population already weakened by an earthquake. Plague is now seldom fatal—modern antibiotics like streptomycin, gentamicin, or tetracycline are effective against it, as long as it is caught in the first 18 hours.[12] Even so, 4 of the 7 people infected in the United States from 1950 to 2006 died. Another death was added in November 2007, when a wildlife biologist at Grand Canyon National Park, Eric York, died of the plague; some 30 people he had contacted were given antibiotics as a precaution, and none of them sickened. York had had exposure to both rodents and mountain lions.

> I had a little bird,
> Its name was Enza.
> I opened the window,
> And in-flu-enza.
> —Children's skipping chant, 1918

How diseases spread among humans depends on population density, on social mobility, and to a lesser degree on the manner of living. Many diseases spread most easily when there is a large reservoir of possible victims. Well, people we have lots of; in many parts of the world, population densities have never been higher. And in many of those high-population reservoirs, people are once again—or still—living cheek by jowl with domesticated animals. Including birds.

Which brings us to the flu virus, "an agile little bastard," as someone from the U.S. Centers for Disease Control once put it.

Viruses seem to have their own rhythms. Some deadly ones like the filoviruses seem to have structures strongly resistant to mutation. Others, like the AIDS retrovirus, mutate at an alarming rate, which makes combating them very difficult. Viruses like it give researchers sleepless nights, because if, say, the AIDS virus ever developed an airborne strain, all our lives would become incrementally more hazardous.

The flu virus, an orthomyxovirus, is one of the most agile of all. It readily spreads by aerosols and seems to mutate as soon as you look at it, which is why this year's flu vaccines will be no use against next year's strain. Under a microscope, the virus seems to have "spikes" sticking out from all sides; they change from year to year due to what is called antigenic drift, giving a new strain each time. Most of the mutations produce illnesses that are, while not exactly benign, at least not fatal. But sometimes they become deadly.

In 1918, a rare genetic shift of the flu virus, or Influenza A, came from China, probably in a crossover from birds, as avian flu. Influenza A viruses occur naturally among birds, though here too there are subtypes, because of changes in, and combinations of, certain proteins[13] on the virus's surface (those "spikes"). Infected birds shed the virus in their saliva, nasal secretions, and feces, passing it on to domesticated birds through direct contact or contact with contaminated surfaces, such as cages. The mortality rate among infected domesticated birds is 90 to 100 percent, death coming within forty-eight hours.[14] The recombination of the surface proteins in the 1918 human variant was novel to everyone. No one was immune. And the strain was horribly virulent.

The illness that resulted was at first called the Spanish Flu, because that's where it first came to medical attention—it killed eight million people there in a month—although there had been earlier outbreaks, notably in military camps in the United States, especially Kansas. The pandemic it caused killed more people than World War I itself. (For a while it was thought to *be* a part of the war and was blamed on a German biological warfare agent, until it was seen that the Germans were dying as fast as anyone else; in some of the trenches, near the end, there was no one left able to lift and fire a weapon.) The death toll in a single year was probably around forty million people, which makes it much deadlier than the Black Plague, which took four years to kill about as many people. Before it ran its course, it had infected a fifth of the global population. Unusually, it was most deadly for otherwise healthy young people between the ages of about fifteen and forty. The mortality rate was 2.5 percent, still far below the filoviruses, but much higher than the normal flu strains, which kill only 0.1 percent of those infected. No part of the world escaped.

The Armistice on November 11 only made things worse—it set off huge celebratory parties as crowds gathered in the streets, the worst possible thing to do, from a public health point of view.

Throughout Europe and America bodies piled up, with not enough workers to bury them. Some towns required a signed certificate of health before outsiders could enter; railways refused passengers without a similar certificate. Funerals were perfunctory, limited to a legal fifteen minutes.

The disease struck with ferocious speed. People died within hours; they'd set off for work, feel ill along the way, and be dead by lunchtime. Four women played bridge together one night, into the small hours; three were dead by dawn.[15] They'd died gasping for air, their airways chocked with a blood-tinged froth that gushed from their noses and mouths.

Influenza A is a stage 4 disease, which means that it crosses readily from animals, birds in this case, to humans and can survive and even thrive in very long cycles of human-to-human contact. But it is more complicated than that. Not all strains do cross from birds. Some that do don't last. Many that do are not virulent. Some that are virulent become less so when they cross to humans. Some become even more virulent. Others can infect humans directly from birds, but not one human from another. Still others set off human pandemics. And the virus can change, and change fast, from one such form to another.

Somewhere around 2003, a new strain, H5N1, appeared in birds in China and elsewhere in Asia. It was a horrid little thing, as dangerous in its way as the Spanish flu of 1918, and maybe more so. Science was holding its breath: as of 2007, it was spreading along the flight paths of migratory birds, but it had yet to transform itself from stage 3 to stage 4. It could infect humans, but only those in direct contact with infected birds—a few hundred had been so infected by 2007, with a dismayingly high mortality rate of 50 percent or greater. A growing number of human cases started appearing in Azerbaijan, Cambodia, China, Djibouti, Egypt, Indonesia, Iraq, Thailand, Turkey, and Vietnam. Infected birds started appearing in western Europe. In November 2007 there was a flurry of alarm as the virus was found rampant in a British turkey farm on the Norfolk-Suffolk border. More than 5,000 turkeys, 400 geese, and 1,000 ducks had to be slaughtered, and health authorities were hoping it would be enough.

If H5N1 mutated just a little—a tiny bit of protein change—and crossed over to stage 4, that would mean real trouble. A study published in the journal *The Lancet* in December 2006 calculated that a 1918-style virus could kill as many as 60 million people or more, a number reached by applying the historical death rate to modern population and mobility data. The researchers, led by Chris Murray of Harvard, calculated a death toll of 51 million to 81 million, with a median of 62. Murray said later he found the number surprisingly high: "I had expected a number of between 15 and 20 million." Around 90 percent of the deaths would occur in the developing world, the study suggested—population density, nutrition, and immune status would all have an effect. Recent flu pandemics, caused by a viral form much less lethal, killed 2 million in 1957 and 1 million in 1968.

Well, the numbers are "interesting," as WHO's Dr. Keiji Fukuda put it, "but not important." What is important is this—what can be done to head off a new pandemic?[16]

PART FOUR

What Is to Be Done?

FIFTEEN

Making Things Worse:
Acts of God and Acts of Man

The world has grown old. The rainfall and sun's warmth are both diminish-
ing, the metals are nearly exhausted. . . .
—Cyprian, 250 AD

I t should be easy enough to distinguish between natural calamities
(what the insurance industry is pleased to call "Acts of God") and
those that are caused by human action or inaction. But in practice, it
is not so simple. The two kinds of calamities often intersect and overlap. As this book has made clear, we occupy a perilously thin habitable
layer on a vulnerable and unstable small planet in a hostile cosmic
neighborhood, prone to shaking, explosions, and poisonous exhala-
tions. Unfortunately, as a species, we tend to aggravate these stresses.
We overpopulate our little planet and re-engineer its ecosystems for
our own convenience. We pollute our environment. We colonize less
and less hospitable places on the Earth. True, we're making life more
pleasant for many of us—for millions of us, in fact. But, while it is not
always easy to assign cause-and-effect, it's hard to argue that we are
making the planet a safer place for anyone, even for those millions who
are currently doing well.

It makes no sense to increase the numbers of people living in those areas prone to flooding (the plains of Bangladesh) or to earthquakes (California) or to mudslides (Guatemala) or to hurricanes (the barrier islands of the Carolinas). But we do it anyway. Are the consequent calamities then entirely "natural"?

When heavy rains set off a mudslide in Honduras that buries three villages that weren't there a few years ago, villages built on a steep hill slope in a region of torrential rain, built there because there was no other land available—is this a natural calamity? When a cholera epidemic breaks out in Brazzaville, Congo, in a mega-slum of more than a million people, a slum in which families have even colonized old car wrecks because they have no money for anything else, a slum that has no sanitation systems whatever—in what way is this calamity natural? When a hurricane slams through a beachfront community in Texas, a community built just above the high-tide level on sandy soil, a community cheaply built with over-skinny lumber, a community whose houses are then flattened by the wind and the storm surge, is this disaster entirely natural? When the Rhine River floods and attacks the Netherlands from behind, is this flood a natural calamity? Even though the river no longer meanders through the countryside but careens through a man-made channel eight meters deeper than it was, a river that no longer has wetlands to absorb its overflow? When Vesuvius erupts and buries villages built on the volcanic tuff at its very feet—how can their deaths count as entirely natural? When a neighborhood in San Francisco collapses in an earthquake, and it turns out it was built on reclaimed land that was largely sand, is this natural? When a traveler with a hemorrhagic fever boards a plane and goes home via four airports in four countries, traveling therefore in a confined space using recycled air together with more than a thousand people, can we blame the consequent outbreak on natural causes?

An instructive illustration of the increasingly blurred boundary between natural and man-made catastrophes was the series of court cases that wound their way through the Louisiana justice system in the aftermath of the disaster of Hurricane Katrina. Those property owners who had any insurance at all found that their policies covered most of the damage done by the wind but seldom allowed claims for water damage or the damage caused by the floods that followed the failed levees. The plaintiffs argued that the flooding was not at all an Act of God, as the insurance companies claimed (thereby shifting the onus for compensation, under disaster legislation, to the federal government) but was on the contrary partly human-caused—arguing that anthropogenic global warming contributed to the storm's fury and that its damages were exaggerated by human inactions, such as not building strong enough levees. Therefore, the claims said, there was fault to be assigned and damages to be claimed. In August 2007 the judge hearing the case finessed the answer in the insurance companies' favor. In the end, he said, it didn't matter in law who or what caused the disaster. All that mattered is that flooding was excluded from compensation.

And of course a larger point: why build a city (or a whole country, like the Netherlands) that is below sea level in the first place?

A re we making things worse? Of course we are. Yes, we are doing much that is good. We are making individual lives better, and even safer, through scientific and medical advances, and globalization has lifted millions of people out of poverty. But we also make whatever calamities do happen worse, in a variety of creative ways: by producing too many people in the first place; by believing in the chimera called economic growth; by anthropogenic climate change; by ecological degradation; by scientific hubris (with its risk of nuclear winter, runaway genetic drift); through pollution and the development of chemicals with unknown consequences; through the weaponization of natural toxins (biological warfare), and many others. It's worth reviewing some of these. Because if we make things worse, we can also make them better.

If we can figure out how.

1. The shibboleth called growth: too many people, too much stuff

In human history, numberless societies have collapsed through ecological degradation. Some of those collapses were no doubt accelerated by purely natural events, such as climate change and desertification, but they collapsed in great part because their citizens were unable to predict change or cope with it when it happened. The great iron-smelting and manufacturing centers of the Middle Niger in North Africa, around the region of Mema, collapsed around the end of the thirteenth century for that very reason—climate change, exaggerated and exacerbated by the destruction of the natural forests for fuel. Easter Island is the more famous example. "Take a few people," Ernest Zebrowski once wrote in an uncustomarily cynical mood, "put them on a remote but beautiful island where they have no practical ways of escaping, let them multiply for fifty generations, and their children eventually outstrip the carrying capacity of their homeland, then turn on each other and destroy all that has been accomplished in generations past."[1] He and Jared Diamond were both using Easter Island as a metaphor for the human species and its occupation of a "remote but beautiful planet, from which they had no practical means of escaping," presumably predicting the same dismal results.

Our species' capacity for being fruitful and going forth to multiply has long ago lost any evolutionary justification. That the United Nations, in its 2007 population estimates, could forecast a population by 2050 of "only" 9.2 billion, and base that on a presumed but unproven future drop in the birth rate in less developed countries, is a sign of how far things have gone. That hoped-for decline in fertility is simply assumed, based on what has already transpired in developed countries, but no proof is adduced. If for some reason it doesn't materialize—if underdeveloped countries continue to manufacture people the way they do now, at a rate of 4.6 children per woman, the population of those regions alone would increase to 10.6 billion instead of 7.9 billion—in other words, without a sharp reduction in the current fertility rate, the world population would increase by twice as many people as were alive in 1950.

Curiously, in all the discussions of remedies for global warming, of how to prevent climate change, population is hardly ever mentioned. The Kyoto treaty ignored it; *An Inconvenient Truth* paid it hardly any mind; remedies from carbon trading, carbon sequestering, alternative fuels, the whole notion of sustainability and the rest—none of them

Human Population History

World population estimate 10/24/2007 at 13.10 GMT:
6,626,665,535

It took all of human history until 1830 for the world population to reach one billion. To reach the second billion took 100 years, the third took 30 years, the fourth 15 years, the fifth a mere 12 years. The "explosion" in human population coincides with the invention of the first mechanical steam engine.

In millions (figures prior to 1850 AD are estimates):

40,000 years ago:	2.5
10,000 years ago:	10
2,000 years ago:	250
1750:	790
1800:	980
1830:	1000
1850:	1262
1900:	1650
1930:	2050
1950:	2520
1960:	3000
1974:	4000
1986:	5000
1999:	5978
2050:	8909

has a prayer of ever working unless the numbers stop increasing. And yet the problem remains the great unmentionable.

It needs to be said, explicitly: the population increase now projected by 2050 would mean that if we could reduce the global carbon footprint by an unlikely 50 percent (with all the efficiency, conservation, and considerable political cooperation that would take), it would all be for naught. The increase in people would simply wipe it out.

When Paul Ehrlich wrote his blockbuster *The Population Bomb* in 1968, he was greatly derided, pilloried in the popular press as a gloomster, a neo-Malthusian; his notions were overtaken, it seems, by the Green Revolution, which proved that we could, after all, feed the world, and the many dire consequences he predicted failed to come to pass; in turn, the fads his book set off, like the Zero Population Growth movement, faded into obscurity.

It didn't much help that the imminent famines he predicted failed to materialize, but what made his book so much less palatable was his strident tone and the dismayingly drastic remedies he urged on governments everywhere: "Our position requires that we take immediate action at home and promote effective action worldwide. We must have population control at home, hopefully through changes in our value system, but by compulsion if voluntary methods fail." This "compulsion" included adding contraceptives to all foods and to drinking water. Talk about the nanny state. . . . In any case, in the developed world, the population seemed to be stabilizing all by itself, without any outside intervention.

But here's the thing: the population bomb did go off, after all. It just exploded more slowly than Ehrlich had forecast. We're feeling the effects of the detonation now. The deleterious effects of the Green Revolution—a biped whose one leg is irrigation and the other chemical intervention—are already obvious, as soils turn saline and unproductive, and chemicals leach into groundwater. Global warming and climate change, pollution, soil and water degradation, pandemics—they are all to some degree caused by the human population explosion.

James Martin, in *The Meaning of the 21st Century*, put it this way: "From 1950 to 2000, the medical profession found ways to eradicate some dreadful diseases and to keep people from dying . . . this was one of the great achievements of its time. . . . [But] excessive population growth leads to poverty, starvation, disease, squalor, unemployment, pollution, social violence and war."[2] It is a classic illustration of the law of unintended consequences.

You don't have to agree with this, or with its implied draconian remedies, to know that human growth remains a massive problem. Put it this way: by the time you finish this sentence, another seven people have been added to the world population. . . . The world is increasing by the equivalent of a Britain every year; Egypt is a country famously thrifty with its water resources, but they are already using all the water available and there are a million new Egyptians every nine months; Niger, one of the poorest countries on the planet and already ecologically stressed, had a population of 15 million in 2007 and could hit 80 million by 2050, without drastic reductions in fertility; Afghanistan could grow from 30 million to 82 million.[3]

"Family planning," as contraception is usually called, has been condemned by some feminist groups as unnecessarily coercive; the U.S. government of George W. Bush reduced funding for contraception on the grounds that it encouraged promiscuity; and the Catholic Church obdurately refuses to think about the issue at all, contenting itself with forbidding its billion adherents even to contemplate it.

In 1994 the UN International Conference on Population and Development acknowledged the global need to slow population growth but could suggest no realistic way of achieving it. Instead, in the wonderfully platitudinous language of such international conferences, it suggested a basket of initiatives, some of which would reduce population, and others, equally necessary and fruitful, that would increase it. They included aiming for gender equality; eradicating poverty and hunger; achieving universal primary education; reducing child mortality; improving maternal health; combating AIDS, malaria, and other diseases; and ensuring environmental sustainability.

On the other side of this fruitless debate are the ecological "activists," as they like to be called, who propose that the Earth can support a population only of somewhere between 2 and 3 billion, maybe as low as 500 million. The more radical among them even point with gloomy relish to the prospect of a global pandemic as a corrective, some sort of necessary Gaia's Revenge. You can even find a few who argue that the world would be better without any humans at all, although "better" for whom or for what is never quite specified.

Meanwhile in my country, Canada, and in my little province, Nova Scotia, politicians were fretting in 2007 about shrinking population levels. This was perceived to be a problem, because without population growth you have a hard time sustaining economic growth. And without economic growth, you are a failure.

In the academies, and in the national councils advising our governments, people were wrestling with the oxymoron called sustainable
growth. Probably Easter Islanders felt the same way.

The rhetoric and literature of the climate change debate (and of
economics generally) are filled with the terms "sustainable
development" and "sustainable growth." The two terms are often used
as synonyms, as though they were the same thing, though they are
really quite different. The word "growth" is the real problem. Growth
is now simply assumed to be a good thing, as though it were also syn-
onymous with "getting better"; in this way, it has taken on an almost
religious connotation of ultimate goodness.

It's true that orthodox economists admit the theoretical existence
of a point in time where growth becomes less than optimal—where
increasing marginal costs become greater than increasing marginal
benefits, in elementary economic theory. But it always seems to be far
in the future; meanwhile, something will surely come along to save
us, as the Green Revolution did. This touching belief is no different
in kind from the equally touching faiths to be found in, say, Christian
doctrine (where paradise is a desirable goal but always somehow in
mankind's future) or in Marxist doctrine (where "communism" is also
a desirable goal that will inevitably appear far in the future when
other objective criteria have been met).

Marty Hoffert, a physics professor at the New York University, puts
it this way: "There's an argument that our civilization can continue to
exist with the present number of people and the present kind of high
technology, through conservation. I see that argument as similar to a
man in a sealed room with a limited amount of oxygen. If he breathes
more slowly he'll be able to live longer, but what he really needs is to
get out of the room."[4]

Again, you don't have to subscribe to this gloomy view to know
that constant growth is a problem. Our economy is predicated on
growth. So is scientific endeavor. In fact, our whole society is organized in a way that depends on constant growth; we're either in growth
mode or in recession. We have to grow to survive. Companies that

don't grow are penalized and bankrupted. The notion of a steady-state economy simply doesn't exist in current thinking.

You can find examples in almost any daily newspaper. China's GNP grew 10 percent last year—good! So did India's—good! Asian tiger—good! America grew 4.6 percent—not quite so good! Belgium grew 1.3 percent—terrible! Canada is forecast to grow more slowly next year—a warning flag for investors. Companies must grow or face the wrath of stock analysts. Dell computers grew only marginally—stock plummets. Apple's iPod and Mac sales grew by 23 percent—stock doubles in value. Firms that don't grow fire their CEOs (albeit with golden parachutes) or become takeover targets. A recession is defined as two successive quarters with zero or negative growth—and a recession is the economic equivalent of perdition. Recessions cause falling employment, business failures, plunging stock markets, human misery.

If really pushed to deal with the question of continuous growth in a finite world and a closed ecology, orthodox theory will generally insist that it will be accompanied by equally continuous technological and scientific development that will continually diminish the effect on the physical environment—in other words, growth can be stripped of its energy requirements and need for natural resources. The contrarian economist Herman Daly of the World Bank sardonically calls this the "angelizing" of economic activity, where economic activity becomes purely ethereal, not using any resources at all. The energy sector is a case in point. It is agreed, though not by all, that the era of "peak oil" may be upon us or imminent, but economists remain unworried. The crisis, if it exists, will be solved by increasingly sophisticated extraction technologies, and then by alternative power sources developed by engineers—either a massive network of breeder reactors, or some basket of alternative technologies (wind, solar, tidal power, beams from space), or the development of nuclear fusion, or some unlooked-for and unheralded technology that will have little environmental impact. It is assumed that the resources used by these activities can be reduced to something approaching zero so we can "grow" them without increasing their environmental impact. Here's a pristine sample of this thinking, from Ray Kurzweil, author of, among other titles, *The Singularity Is*

Near: When Humans Transcend Biology: "[To solve the greenhouse problem] . . . we need to capture only 1 per cent of 1 per cent of the sunlight to meet all our energy needs (3 per cent of 1 per cent by 2025) and nanoengineered solar panels and fuel cells will be able to do this, thereby meeting our energy needs in the 2020s with clean and renewable methods. . . . As a result, poverty and pollution will decline and ultimately vanish, despite growth of the biological population. . . ."[5] But the impact of these technologies cannot be reduced to nothing. As Daly argues, sooner or later, angelizing efforts will reach a point at which the amount of energy and other resources saved by further improvements in technology become minimal. This will make further significant increases in the volume of production impossible without causing additional environmental damage.

The week I was writing this passage, the toymaker Mattel recalled millions of cheap Chinese-made plastic toys because they were contaminated with lead. In the hullabaloo that followed, the preferred solution seemed to be to transfer the manufacturing to American workers. It never occurred to anyone, even the critics of rampant globalization, that maybe kids could do with fewer toys or perhaps use their inborn creativity to make a few playthings themselves.

Herman Daly calls the idea of sustainable growth an impossibility theorem.

E xcept in the cant of the self-improvement industry, where growth seems to mean making you a more fulfilled person, the verb "to grow" means to get bigger, not better. The word "development," on the other hand, means improving quality, not quantity, and implies fulfilling or realizing potentials. If sustainable growth is impossible, sustainable development is at least plausible. Our planet's closed ecosystem cannot grow, but it can develop.[6]

Sustainable development, as first defined by the Brundtland Commission (the 1983 World Commission on Environment and Development, under the leadership of Gro Harlem Brundtland) and then by Maurice Strong's Earth Summit in Rio in 1992, means just that: balancing human needs while protecting the natural environment so those needs can be met not just in the present but the indefinite future. In 2005 the UN's World Summit Outcome document amplified the point, suggesting that sustainable development includes the "interdependent and mutually reinforcing pillars of economic development, social development, and environmental protection." (It's true that the document then got a little woollier, suggesting that "cultural diversity is as necessary for humankind as biodiversity is for nature" and once again confused the two terms, going on to say that "development [should not be] understood simply in terms of economic growth, but also as a means to achieve a more satisfactory intellectual, emotional, moral and spiritual existence.")

2. Anthropogenic climate change and global warming

The first "global warming forecast" was issued in August 2007, by British scientists. The prediction: a plateau for two years, then a sharp climb through 2014.[7]

It's true that skeptics still exist, and not just on the wilder shores of the paranoid seas. There is room for skepticism because climate science is fiendishly complicated and still ill understood and also because in their enthusiasm to make the case, those convinced of warming have perpetrated many considerable exaggerations (sea levels have not been rising dramatically around the Maldives; not all glaciers are melting; Kilimanjaro's ice cap is melting because of deforestation; evidence for the increase of extreme weather events is very thin; the increased hurricanes in the Atlantic are probably due to a natural oscillation; tropical cyclones have actually been decreasing in number globally).

Doubtful declarations aside, the evidence for global warming is much greater than the evidence against. If there is not complete unanimity, there is at least a widely accepted consensus.

What causes it is still an urgent question, not yet satisfactorily or fully answered. What its effects will be is also of more than just academic interest. And so is the question of just what we can do about it—the real policy questions are whether we should be concentrating on preventive or remedial strategies, a point that will be dealt with in the remaining chapters.

S uspicions about anthropogenic global warming are not recent and
nor is human-caused climate change itself. The long history of
slash-and-burn agriculture, and the systematic hunting to extinction of
continental wildlife, leaves very little room for comforting myths about
a lost Edenic past. William Ruddiman, author of a number of standard
climate science textbooks, contends that our species has been changing the climate not just since the industrial revolution but far longer—
perhaps as long ago as eight thousand years. His work has been widely
disputed, but the most interesting of his data show that steep declines
in the atmospheric carbon signature (as recorded in ice core data) happened immediately after the great plague pandemics of history, when
millions of acres of farmland were abandoned and returned to forest,
which seems to suggest that climate is more mutable than previously
thought. Which can be interpreted as good news: it can be fixed.

The contemporary climate change debate dates back to 1956,
when Gilbert Plass of Johns Hopkins warned that CO_2 controlled the
climate, and further that it is an unbalanced control, lacking any stability: "There is no possible stable state for the climate." He said that
humans had already increased CO_2, and that warming would surely
follow.[8] In 1957, Roger Revelle and Hans Suess published their find-
ings on CO_2 levels in the atmosphere, confirming earlier suspicions.
"Human beings are carrying out a large scale geophysical experiment
of a kind that could not have happened in the past nor be reproduced
in the future," they wrote. "Within a few centuries we are returning
to the atmosphere and oceans the concentrated organic carbon stored
in sedimentary rocks over hundreds of millions of years. . . ."

The newly minted Nobel laureates called the International Panel
on Climate Change (IPCC), from which the public derives most of
its information about global warming, has long ago accepted that the
Greenhouse Effect, a term first coined in 1827 by the French mathematician Jean-Baptiste Fourier, is both real and, on the whole, beneficial (were there no greenhouse effect, the average global temperature
would be somewhere around minus 0.4 degrees Fahrenheit). Water
vapor is the most abundant greenhouse gas, followed by carbon dioxide, methane, and other trace gases. CO_2 levels held pretty steady,
around 280 parts per million, for the thousand years before 1800. Since

then, as industrialization really got under way, atmospheric concentrations of CO_2 began to rise. Today they are around 370 parts per million. Some people think this isn't very much—after all, CO_2 is only a very small component of air, just a trace gas really, and an increase of 90 parts per million, if visualized as distance, would be 1 centimeter in 100 meters. On the other hand, it does represent a 30 percent increase, and an increase on one side of a precarious balance, at that. Most Earth scientists believe that with the still-expanding fossil fuel infrastructure, it may be impractical to avoid 440 parts per million, a significant increase from present levels. The IPCC warns it could even reach 1,260 ppm, 350 percent above the pre-industrial concentration.

Some uncertainty still exists about the magnitude of the changes under way. Some IPCC critics have suggested the report underplays the cooling effect of pollutant aerosols, which reflect sunlight and increase the reflectivity of clouds.[9] Others, by contrast, suggest that the changes are happening faster than IPCC estimates. Still, the IPCC and its critics are nevertheless "certain" that emissions from human activities are enhancing the natural greenhouse effect, resulting on average in an additional warming of the Earth's surface. Carbon dioxide, though a less potent greenhouse gas than methane, or even water vapor, has been responsible for at least half of the observed increase. Based on business-as-usual scenarios, the globe would continue to warm, by as much as 6 degrees before century's end, and perhaps more; sea levels will consequently rise—possibly by as much as 33 feet per degree, a startling number. "We are risking," the 2005 IPCC report said, "the ability of the human race to survive." The IPCC report issued in November 2007, just before a major international meeting on climate change in Bali, came close to suggesting it might already be too late to change, and urged policy makers to step up action on adapting to, instead of heading off, accelerated global warming. A major die-off of thousands of species is imminent, the report added.

A study by James Hansen of NASA's Institute for Space Studies and other scientists has concluded that the Earth's "energy imbalance," the net heat gain over heat loss, is almost one watt per square meter of Earth surface (enough, the authors say, that if the net gain were maintained for ten thousand years it would be enough to boil the oceans). Gloomily, they go on to suggest that rapid climate change might take less than a century, while it would take at least a century to change our ways even if we started now, "implying the possibility of a system [already] out of our control."[10] Hansen is one of the pessimists. He thinks the IPCC reports are far too conservative and that changes are coming much faster than suggested.

No one is suggesting that natural forces are not also at work. Natural climate cycles might even cause blips in the warming, which in turn might be interpreted as contradicting the data. For example, the journal *Science*, reporting on a meeting of the American Geophysical Union in San Francisco in December 2006, reported that the High Arctic, which it called "the poster child for global warming," may very well get one or maybe more temporary reprieves: "The recent warming has a natural component, and all the models show frequent increases and reversals, so there is bound to be one or two in the next decade or so. The region may even get cooler until the next natural swing to the warm side once again reinforces the greenhouse [effect]." James Overland of the U.S. NOAA Pacific Marine Laboratory, in Seattle, says the models predict a five-year cold period by 2010, and "then people would say, aha, we don't have global warming." Some scientists fretted that the public will have trouble grasping global warming that regionally and temporarily goes away.[11]

The rise in historical CO_2 levels since 1760 and the temperature increase since 1875, the first credible measurements, are closely correlated. In the 1940s the Earth began cooling down, a trend that lasted until the 1970s, when it began warming up again. The explanation is that the effects of the increased CO_2 were masked by the increasing pollution (soot, smog), which had a cooling effect. Cleaning up the polluted air has freed the increased CO_2 levels to further warm the atmosphere, which it has been doing steadily for the last thirty years.

One of the last refuges for greenhouse skeptics was the notion that the current warming trend has been produced not by human activities but either by an increase in solar radiation, or by a decrease in the flux of galactic cosmic rays. But a new study by Mike Lockwood, a physicist at the Rutherford Appleton Laboratory in Britain, correlated global temperatures with the eleven-year solar cycle and found that "trends in solar irradiance, sunspot number and cosmic ray intensity have all been in the opposite direction to that required to explain global warming."[12]

I don't have to flog this point. It is abundantly clear that human-induced global warming is likely to make a lot of things worse and only a few things better. Changes for the worse include increasing aridification and even desertification in already dry areas such as the American southwest, North Africa, and north China; longer and more severe droughts, leading to famines; flooding of low-lying countryside and coastal cities; more intense rainfall in mid- to high-latitudes, and more flooding; increasingly intense tropical storms; and the collapse of traditional agricultural practices.

3. Scientific hubris

Nothing could be more insane, surely, than MAD, the Cold War doctrine of Mutually Assured Destruction.

One of the best places to get a sense of the scale of the madness is 722 feet below the swanky Greenbrier resort in the Allegheny hills of West Virginia, in the echoing caverns of the bunker built by the Eisenhower administration in the 1950s to house the United States government during, and presumably after, a nuclear attack by the "Commies" in Russia. The whole complex was decommissioned by the federal government only after *The Washington Post* uncovered its existence in 1992. Until then it had, at least in theory, been a secret, although recently released Russian files indicate that the Soviet Union had a pretty good idea it was there all along, such secrets being very much to the Russian way of thinking.

In the paranoia of the time, the Eisenhower administration did a deal with the owners of the Greenbrier, to which people from Washington had been coming since 1788 to "take the waters" in the sulfur springs of the area. As a quid pro quo for allowing themselves to be used to disguise this mooted Emergency Relocation Center, the Greenbrier acquired several new convention and meetings rooms, aboveground but with hidden blast doors, that they would turn over to the fleeing Congress when the time came. Another 112,537 square feet were excavated deep underground. Construction was finished in 1962, and for the next thirty years the resort's owners kept mum as instructed, while agreeing to turn over the whole facility to whatever remnants of the legislative branch survived the nuclear emergency (the executive branch and the president were to be conveyed else-where—separation of power was to be maintained at all costs).

A train line was built to White Sulphur Springs, the community in which the Greenbrier is located, so the federal government could be easily transported without causing traffic jams, assuming the Russians sportingly gave everyone enough warning. An airport strip was laid for those in more of a hurry.

The whole complex was run by federal agents disguised as employees of a dummy company named Forsythe Associates, an audio-visual company. Phone calls were routed through the hotel switchboard to maintain the fiction.

Walking through the complex now—which you can do since the Greenbrier resort operates tours for its guests (and rents some of the facilities to corporations for "themed" events such as James Bond or M★A★S★H evenings)—is quite surreal. Hundreds of feet of now empty storage spaces, a hospital, a crematorium for biologically haz-ardous waste (i.e., dead humans), decontamination chambers, a gen-erator and power supply, water filtration systems, dormitories big enough for all members of Congress and their staffs, VIP suites, kitchens, broadcast studios with painted four-seasons backdrops of the U.S. Capitol so the congressmen's constituents, should any have sur-vived, could be reassured that their government was in place and functioning. To cheer everybody up, the dining room featured false windows with wooden "frames," beyond which the sequestered legislators could "see" bucolic country scenes.

The paranoia had a very long reach, even to Nova Scotia, speak-ing of bucolic scenes. The house we bought there had been built in the 1970s by a survivalist from Tennessee who had become convinced that a nuclear holocaust was going to devastate the United States. As a consequence, he took refuge in Nova Scotia, although why he thought this little jurisdiction would be spared the fallout is unclear. When he returned to the United States in the 1990s, presumably convinced that a stake had been driven through communism's heart and it was now safe to go home, he left behind a hefty volume of a thousand pages or so that turned out to be the official U.S. government's advice to citizens on how to reconstitute civil power after the collapse and the presumed annihilation of federal authority. It made illuminating reading. There were chapters that were essentially primers on reconstituting a secure food supply for local communities, on power generation and communications technology, on organizing elections and vetting candidates, on dealing with vigilantes (mostly by co-option and a "command structure") and looters (by shooting them). At the end, as an appendix, was the U.S. constitution. All citizens were urged to keep the manual to hand, perhaps in their fallout shelters along with whatever emergency supplies they had stocked there. (In the case of our house, the owner had walled up a large barrel of dried beans.)

You find all this ridiculous, akin to fears of the black helicopters of the UN coming over the hills to sap the nation's bodily fluids? Then consider this: On October 21, 1998, President Bill Clinton signed Presidential Decision Directive 67, which was titled "Enduring Constitutional Government and Continuity of Government Operations." The reason was not so much nuclear winter as societal collapse caused by terrorism, pandemic, or some other disaster (the directive was open-ended in its gloom); the rationale was to make sure government survived, partly by laying out plans for the succession but also to provide bolt-holes for fleeing legislators. Then, as *The Washington Post* reported, George W. Bush after 9/11 set up a shadow government of about a hundred senior civilian managers to live and work outside Washington on a rotating basis, to ensure the continuity of national security. Since then, a program once focused on presidential succession and civilian control of U.S. nuclear weapons has been expanded to encompass the entire government.

This is not all bad. Ensuring properly constituted civilian control of weapons of mass destruction can only be a good thing. Because while Saddam Hussein might not have had WMDs, the United States has plenty to go around.

M AD's risks were not just deliberate nuclear assault, though that was bad enough. The highly automated weapons systems and the imperative of speedy action under threat also made the risk of purely accidental nuclear war substantial; accidents could easily have happened either through human carelessness or through equipment malfunction. Just so in June 1980 a defective computer chip at NORAD generated a warning of Soviet attack and a nuclear alert was issued (the command director being dismissed the next day for failing to evaluate the situation correctly): "Seven months earlier, B-52 bombers had been prepared for takeoff, while intercontinental ballistic missile crews started preliminary launch procedures . . . because a wargame tape had accidentally been allowed to control the main warning displays."[13]

It is true that the threat of nuclear winter has receded somewhat, with the dismantling of the Soviet and, to a lesser degree, the American arsenals. But the threat of the deployment of nuclear weapons is once again on the rise. Unemployed Soviet physicists have been keeping their families alive by working for such stable entities as Iran, Pakistan, North Korea. Alarming amounts of plutonium have mysteriously gone missing, and France's fast breeder reactors (neutron reactors that produce more fissile material than they consume) are busily churning out more all the time. India and Pakistan have fought conventional wars against each other since independence sixty years ago, and both are armed with nuclear bombs. Israel has them too. So does (or did) South Africa, although the why of their arsenal remains a mystery—they could have deployed them only against their own people, in a mad fit of self-immolation. Iran is actively seeking a nuclear arsenal and so is the Hermit Kingdom, North Korea. Syria is rumored to have been doing the same, before the Israelis intervened. That terrorist networks have been seeking a nuclear bomb is no secret, nor is the certainty that such a weapon would be deployed if, or rather when, acquired.

Of hatred there is a sufficient global supply, an entirely sufficient megatonnage, and there is no sign it is being decommissioned.

S cientific tinkering has not, obviously, been confined to the nuclear military axis. A lot of other technologies can be turned to mischief too.

Take the bio-technologies. Much of the opposition to—and revulsion against—genetic engineering has been focused on the continuing efforts to create chimeras, genetic bastards, often animals sharing human genes, or humans sharing animal genes, or animals sharing spider genes—the range is endless.

This popular revulsion has been met, by and large, with scientific derision; the governing scientific paradigm is that any research that furthers human knowledge should be permitted and encouraged, and that opposition automatically implies some kind of anti-scientific Luddism, a creature of this or that obscurantist and anti-enlightenment philosophy. After all, none of the popular fears has been borne out: nanotechnology has not produced the "gray goo" that its opponents feared, runaway self-replicating nano-machines a single micron in size capable of turning the entire planet into a homogenized mass (although a fourth nano-generation expected soon will actually mimic mammalian cells, with all the hazards that entails). Exploding thermonuclear or fusion bombs did not ignite atmospheric hydrogen, set the air on fire, or strip away the ozone layer. Plasma and high-energy particle physics has not created black holes, scalar breakdown, or vacuum meta-stability effects, as we saw in the opening chapter. Genetically engineered pathogens (created "to see what they may be like") have not escaped into the atmosphere. The AIDS virus has not been mutated into a creature with an aerosol vector. . . .

None of these things has happened. Yet. We remain alive. Still.

4. The weaponization of calamity

It's pretty obvious that human beings have never flinched from harming others of our species—we seem hard-wired to kill anyone outside our tribal family. What other species would deliberately inflict plague on its fellows? Nor have we been reluctant to employ natural forces to do our work for us.

Peter Gleick, who runs an environmental and resources think tank in California, has collected more than two hundred instances of violence involving that most necessary of resources, water. Very few of these, surprisingly, involved war over water; instead, almost all involved using water as a weapon of war. If you ignore Jehovah's genocidal use of water to kill all humanity except Noah and his family, the oldest known instance dates back to 2500 BC, when Urlama, king of Lagash, diverted water away from canals owned by his rival, the state of Umma.

Many others can be found, all through history to modern times. For example, in 600 BC the Athenian Solon poisoned the water supply of Cirrha during a siege, making the enemy violently ill. As recounted in Chapter 12, the dikes containing the Yellow River, or Huang He, have several times been deliberately breached to drown opposing armies, in each case causing hundreds of thousands of deaths.

Biological agents have also been used, long before anyone knew why they did what they did. Stone-age arrows dipped in animal or human feces have been found. Alexander the Great once won a siege by having his catapults hurl diseased cattle into the defenders' piazza; the Roman empire used animal carcasses to poison enemy wells; Hannibal hurled amphoras of poisonous snakes onto an enemy ship; plague was several times used as a weapon of war in the runup to the Black Plague of the Middle Ages, thus ensuring its even faster spread, a device reprised by the Russians in 1710 against the Swedish invaders in Estonia. Biological warfare came to the New World in the fifteenth century, when the *conquistador* Pizarro gave clothing contaminated with smallpox to Native Americans; Britain's Jeffrey Amherst did the same thing to France's Indian allies, giving them blankets that had been used by smallpox victims. During World War I the German army experimented with anthrax, glanders (rare in nature, and acquired from horses), cholera, and wheat fungus as biological weapons, spreading plague in Russia and Mesopotamia. Although the Geneva Protocol of 1925 was signed by virtually every nation, Japanese forces in China exposed more than three thousand victims to plague, anthrax, syphilis, and other agents. The secret Unit 731 of the Japanese army dropped plague-infected fleas over populated areas of the country under occupation, causing sporadic outbreaks of plague.

The weapons shops of the United States and Soviet Union housed a ghastly menagerie of infectious agents, partly for protective purposes but also to see whether they couldn't be employed as assault weapons. *Yersinia pestis* (the plague bacillus), smallpox, anthrax, Ebola, ricin, botulism, brucellosis, tularemia, and others—experiments were done on them all to see whether they could be mutated and worsened. Both countries developed a form of aerosolized plague, eliminating the need to invoke the flea vector; in 1970, the year the United States theoretically eliminated the program, WHO estimated that if 110 pounds of *Y. pestis* were dropped over a city of 5 million, 150,000 people would likely be affected in the single hour the pathogen would remain viable. Weaponized plague would cause primary pneumonic plague, which does not produce buboes. The first signs of illness would be fever with cough and dyspnea (difficult breathing), sometimes with bloody, watery, or purulent sputum. It would be treatable and—the cheerful side of all this—there would be no need for the conquering forces to decontaminate the area afterwards, since *Y. pestis* cannot live for long outside its host. On the other hand, the flea that carries it is hardy enough; it can survive for up to a year in dung, abandoned rats' nests, or untreated textile bales.

The very ghastliness of bio-weapons helped to deter their use; the flu that killed so many millions in World War I was no respecter of frontiers, nor could it be induced to kill patriotically. No matter one's opinion of the debacle that was the invasion of Iraq in 2002, the notion that biological agents might be in the hands of a creature like Saddam Hussein was enough to silence some early opposition. Fiction has had more sport with this notion than reality, fortunately. That other "agile little bastard," Ebola, has featured in a number of novels, including one by Tom Clancy that turned on an improbable plot by mad environmentalists to return the world to "nature." More chilling was Frank Herbert's story of bio-revenge, *The White Plague*, in which a grieving scientist deliberately engineered a virus that made all women sterile—chilling because in 1993 the salmonella bacillus was similarly genetically altered to cause "a harmless, temporary" intestinal infection that just happened to be a potent spermicide, effectively preventing conception. In that case, it was to be taken deliberately, as a birth control device, but if it had become infectious it would have had catastrophic results. Opponents were written off as anti-scientific cranks.

The nerve gases tabun, sarin, and soman were discovered but not used, as far as anyone knows, during World War II. By 1967 the United States had stockpiled four thousand tons of VX, a toxin so deadly that a drop on the skin could kill. The U.S. army managed in the 1970s to clone a gene for Shiga toxin, which was deadlier still.

Now, new diseases can be produced easily enough by gene splicing and deliberate mutation ("site-directed mutagenesis").

In the new millennium, it is easy enough to set up a bio-weapons lab. Pretty well anyone with a microbiology degree and a few thousand dollars can do so. Delivering it safely is much harder, but it has been attempted, several times. In 1984 the Bhagwan Shree Rajneesh sect (a cobbled-together syncretic spiritual movement led by an Indian-born guru who called himself Osho) deliberately infected salad bars in Oregon (which you would think would be a low-priority target) with salmonella, infecting 751 people, none of them fatally; a few years later a Japanese sect named Aum Shinrikyo sprayed aerosolized anthrax from rooftops in Tokyo, though fortunately their fanaticism was not matched by their expertise; and in 1995 a Minnesota survivalist group was convicted of possessing home-cultured ricin, with which they intended to attack local federal offices. And in 1995, in Ohio, a microbiologist was arrested for fraudulently acquiring *Y. pestis* in the mail—a revelation to a public that had naively thought the stuff must be rather hard to get.[14]

Still, the military being military and politics being politics, not everything is being left to the amateurs. By the end of the Cold War, Russian scientists had stockpiled enough *Y. pestis* to kill several million people; thousands of scientists were at work busily churning out this deadly thing. Neither Russia nor the United States any longer admits to producing biological weapons—except that both are still stockpiling pathogens to study "for purely defensive purposes." (This of course is "pure mind-flannel," as the writer Nelson DeMille once put it. "Bugs don't know the difference between defensive research or an airburst bomb. They don't even know whether they are good bugs or bad.") In 2007 the U.S. government released a short list of sites that were being considered for a new $450-million "bio-security complex" to be run by the Department of Homeland Security. The press release said the new facility would do research on potential bio-weapons (they called it "bio-defence") and replace the fifty-year-old Plum Island Animal Disease Center, off Long Island, hitherto the preeminent manufacturer of noxious diseases, long controversial mostly for being practically within view of some of New York's most affluent summer homes. Left off the short list was Texas A&M University. Two workers had been exposed to "potential bio-weapons" there; to make matters worse, the university had waited a year to report the incident. Even more unnervingly, three vials of the nasty little bacterium called *Brucella* had simply gone missing. It's not just an American problem. In Britain, a similar lab let loose a strain of foot and mouth disease— it simply ran out through a faulty drain system, and thousands of cattle were put at risk.

Were any labs working on an aerosolized version of Ebola? The answer, alas, is yes, even outside the novels of Tom Clancy. It is not known whether they found one. Or how they'd contain it if they ever let it loose or it somehow escaped.

"Proponents insist that there is a clean safety record. That is simply wrong," a critic of bio-weapons labs, microbiologist Richard Ebright of Rutgers University, asserted in October 2007.[15]

Few natural systems have escaped the notice of the weapons makers. In 1996 the Pentagon issued a report by Air Force Chief of Staff General Ronald Fogelman called *Weather as a Force Multiplier;* the report imagined that if they could only learn to fine-tune weather systems on command, there would accrue "tremendous power that could be exploited across the full spectrum of war-fighting environments." That is, they could deliver, quite literally, blitzkrieg on demand. As Ando Arike once put it in *Harper's*, "We may one day come to own the weather only to find that the weather has turned as ugly and rapacious as we."[16]

5. Fouling the nest

Sometimes it is the quirkiest things that catch the attention. It was only when I understood that suntan lotion had become the great polluting threat to the Mediterranean's beaches that I finally understood the overpopulation problem, and it was only when I read in a scientific paper that human dandruff, as well as "skin fragments" (and, curiously, trace amounts of cocaine), had been found in high-altitude atmospheric aerosols that the full extent of human polluting activities finally sank in.

Well, the world is alive to the pollution problem now (in 1998 the Eastern Orthodox Patriarch Bartholomew proclaimed a new class of sins—those against the environment) and in some ways many things have been getting better. Stringent air-quality controls in Europe have decreased sulfur pollution substantially. The ozone layer is being repaired, thanks to an international convention. America's air is cleaner than it was, except that the Bush administration gutted the Clean Air Act and the country is now backsliding. New York's Hudson River once again has a commercial fishery, thanks to one of the Kennedy clan. The Rhine River has fish in it for the first time in thirty years. The Great Lakes are cleaner than they were, though still undrinkable without treatment.

But that is pretty much it for good news, at least as reported. Sulfur dioxide hasn't decreased globally, because a wash of pollution from Asia is more than compensating for the decrease in Europe. For the rest . . . Global winds carrying ozone and carbon monoxide (CO) are jeopardizing agricultural and natural ecosystems worldwide and are having a strong impact on climate. All the studies recognize what is already obvious: that Asian pollutants are beginning to surpass those from North America, a trend that can only continue and accelerate. The increasing concentration of carbon monoxide in the atmosphere worldwide is particularly worrying. The importance of megacities, defined as cities with more than ten million inhabitants, is recognized as a new and critical source of pollutants, especially from burning fuels—by 2001 there were seventeen megacities worldwide. And a final point: pollution from elsewhere is making local conditions worse, pretty well everywhere.

In the late summer of 2002 I had been filming in north China, on the fringes of the Gobi, and had noticed that the air overhead seemed curiously opaque, even milky; there was no blue, even on clear days. This was more evidence of China's inexorable desertification and consequent dust storms. Alas, Chinese efforts to correct the problem may be making it worse—in mandating that marginal land must be used for farmland, they simply encouraged practices that caused the newly plowed soil to blow away. In 2001 NASA had tracked a massive dust storm originating in north China big enough to briefly darken skies and cause hazy sunsets over North America as little as five days later; a month after I left to return home, NASA's satellites once again picked up an immense dust cloud, a mile or so thick, moving eastward over Korea and into the Pacific. Clouds like it seem to be becoming part of the Chinese calendar. Similarly massive dust clouds occurred in 1997, 1998, 2000, 2006, and again in 2007; the Chinese Meteorological Agency counted twenty-three major dust storms in the 1990s, a substantial increase over previous decades. Chinese dust, rich, if that's the right word, in pollutants such as coal-combustion aerosols, ozone, persistent organic pollutants (POPs), and heavy metals such as mercury, has inflicted itself on Korea and Japan for decades; in Korea it is sometimes called "spring's gatecrasher." Chinese dust may have been the origin of an outbreak of foot and mouth disease on Korea's west coast.

In July 2007 the head of China's environmental agency admitted that anger at the country's pollution levels was causing riots, protests, and petitions across the country. A World Bank report issued the same month, only after China tried to have it suppressed, asserted that nearly five hundred thousand people die each year from pollution-related illnesses.

It's not just China. In the summer of 2002, a United Nations Environmental Program (UNEP) study confirmed the existence of yet another hazard, a two-kilometer thick pollution cloud over much of southern Asia. The haze was the result of forest fires, the burning of agricultural waste, dramatic increases in the burning of fossil fuels in vehicles, industries, and power stations and emissions from millions of inefficient cookers burning wood, cow dung, and other biofuels.

The author of the report, Partha Dasgupta, a Cambridge University professor, put it this way: "Disaster is not something for which the poorest have to wait; it is a frequent occurrence."

Indian researchers have studied the high concentrations of black carbon (aka soot) over the Indian Ocean, and traced it back to biofuels, mostly cattle dung, used for cooking fires by millions of people in the subcontinent. Only by changing the way India cooks, the study suggested, could the country help mitigate climate change. They acknowledged that change was less than likely.[17]

In August 2007 there was more depressing news. A study led by Scripps Institution of Oceanography scientist Veerabhadran Ramanathan found that the brown clouds of pollution over South Asia have multiplied solar heating of the lower atmosphere by 50 percent. These "brown clouds" are not just water vapor; they also contain soot, sulfates, nitrates, hundreds of organic compounds, and fly ash from urban, industrial, and agricultural sources. As Dr. Ramanathan put it, "The conventional thinking is that brown clouds have masked as much as 50 percent of the global warming by greenhouse gases through the so-called global dimming. While this is true globally, this study reveals that over southern and eastern Asia, the soot particles in the brown clouds are intensifying the atmospheric warming trend caused by greenhouse gases by as much as 50 percent."[18]

A WHO meeting in Bangkok in August 2007 was given some stark news: each year in southern Asia some 6.6 million people die from what the agency called "environmental factors," mostly dirty air. This would be an astounding 25 percent of all deaths in the region.[19]

It is no better with the other natural resource we all share, water. It is not difficult to find examples. Actually, if you are only halfway looking, it's rather hard to avoid examples. Here is a representative finding, by no means exhaustive:

An extensive survey of the world's rivers found only two fit for drinking in their "wild" state, and those only sporadically. Although some rivers are getting cleaner (among them the St. Lawrence), not a single major American river is any longer safe to drink from without chemical treatment; in Latin America and Africa, less than 2 percent of human waste is treated; more than 80 percent of

China's rivers are so degraded they can't support fish. The Yellow River can't even support irrigation—in 2002 a hundred tons of lettuce washed with river water was withdrawn from the market as unfit for human consumption. China's rivers contain more than 10,000 times the bacterial count of Canada's; Canada is not so great either. Canada still has a major city, a provincial capital, that dumps its sewage directly into the ocean—Victoria, British Columbia. In Vancouver, which does treat its sewage, sort of, one plant alone releases 211 billion quarts of under-treated sewage into the oceans each year. Another provincial city, Halifax, has only treated its sewage for a few years; Canadians, one at a time and in small quantities each, pour more oil (used motor oil) into the country's waterways than was spilled by the Exxon Valdez disaster; Bangladeshi water for almost 80 million people is contaminated with varying degrees of natural arsenic; the same problem exists in north India, the unintended consequence of a well-meaning effort to provide wells to the countryside; some 80 million tons of contaminated sediment pours into the Black Sea every average-flow year; the sea itself has been eutrophied and polluted; a credible estimate is that a child dies somewhere in the world every 6 seconds due to contaminated water or water-vector diseases; more than a billion people have no access to sanitation systems or safe drinking water, things the ancient Romans took for granted, and the Babylonians before them, and the numbers are increasing every year.

W ho, or what, is at fault here? It is a dismal balance sheet, and a good forensic accountant would have no difficulty assigning blame. It's not hard to find corporate fingerprints in many places they shouldn't be, and carelessly ignorant governments are more common than not. Corporate entities most often don't pay their share. They don't pay for the common resources they use (fish, water) and they certainly don't pay for the damage they do. The makers of CFCs paid not a cent to fix the ozone hole they helped cause; fishing companies don't pay for the damage their trawlers do to the sea bottom; oil extraction companies don't even acknowledge the damage they do to the air and the ground and the water; manufacturers do not pay to clean up rivers and aquifers they pollute; and because they don't pay, they don't care. But overall, many of these are problems caused not by the corporate villains and corrupt bureaucrats of Green legend but by far too many people far too often doing quite ordinary things in an everyday way. None of these billions of ordinary actions cause natural calamities. But almost all of them make it much harder to respond to calamities when they occur, sapping resistance and enervating public response. The fault is ours to fix.

SIXTEEN

Making Things Better (i): Mitigating Natural Calamities

What is to be done?
—title of pamphlet by Vladimir Ilyich Lenin

If you're in a car crossing a bridge and the bridge suddenly collapses, pitching you and your vehicle into a ravine, you would be forgiven for classifying the event (provided you survived) as a disaster. But a single bridge collapse is an accident, not a disaster, notwithstanding the dire consequences for those using it at the time. On the other hand, if the bridge collapse was caused by the passing of a major hurricane, which also caused widespread flooding and destruction, then the storm would be a disaster and the bridge collapse would be considered one of the elements that made it so. But if the bridge collapse was caused not by a single storm but by a cataclysm, such as a major volcanic eruption that also spewed billions of tons of volcanic ash into the atmosphere and killed thousands of people, then the bridge collapse would merely be a micro-event in a major calamity.

There is a hierarchy of awful things that could happen, and pitching you into a ravine from a mangled bridge could be a small part of any number of them. For you, the effects of accident, disaster, and cataclysm would be identical. For our species, however, the results range from merely sad through tragic to dire. We need, in short, to range calamities by scale along with their strike probabilities.

As a reminder, the reinsurance giant Munich Re, on whose sober actuaries the company depends for its survival, forecasts that three to five major disasters a year will each kill more than 50,000 people. Nonetheless, with some exceptions—some of them outlined in previous chapters—the global death toll from floods, droughts, earthquakes, tsunamis, volcanoes, and cyclones is still quite small—perhaps 80,000 is an average year. Some years the toll is higher—in 2005, with the South Asian tsunami and the Kashmiri earthquake, the toll reached almost 400,000 people. But three times that number died on the roads in the same year, and more than twenty times as many, most of them children, died from drinking contaminated water and from water-vector diseases.

But of course, as Quirin Schiermeier put it in an essay in the journal *Nature*, this rather misses the point.

> Catastrophes are not average; they are the great exceptions. Most of us have seen road traffic accidents, but few have witnessed a natural disaster. They are ruptures . . . that change cities, countries and whole regions forever. They can even change patterns of thought. The Lisbon earthquake of 1755 and its associated tsunami, which struck while the city's churches were celebrating All Saints Day, shook the faith of millions and altered that century's intellectual landscape irrevocably.[1]

That disasters are rare, the great exceptions, makes planning for them difficult. It is easy to overlook something that might happen only once in a generation, and sometimes less frequently than that. Historical memories are short, particularly in the face of population and development pressures. An example of where memory fades even in the short term is in the frequency of Atlantic hurricanes. In the 1970s and 1980s, fewer hurricanes struck the American southeast than in previous decades—and all too soon, this came to seem the norm, despite recurrent expert warnings. Resort hotels, condominiums, and entire subdivisions were built on unstable soil less than a three feet above sea level. Many of them were cheaply built without thought for high winds or storm surges. When the normal hurricane frequency cycle resumed in the 1990s, they were due, in the words of the director of the U.S. Hurricane Center in Miami, Bob Sheets, "for some very unpleasant surprises." And sure enough, soon it was accepted that hurricanes were "getting more frequent and more severe," but mostly because they affected ever more people.

Longer-term cycles are incrementally harder to predict and plan for. A comet appeared in the sky in 1066, the same year Britain fell to Norman invaders; it was easy enough to speculate that kingdoms fall when comets appear—after all, as Ernest Zebrowski points out, it was observationally true. Of course, to need validation, such a theory needs more data—but such data may only be to hand long after the original observers are dead. "Which is why the fledgling science of astronomy gave rise to the mystical pseudoscience of astrology."[2]

Population increases and ecological degradation set the stage for more frequent, more powerful, and more destructive disasters. Vulnerable urban populations, especially in the developing world, are set to double by 2030, as are coastal populations everywhere. Communities already disempowered by economic and ecological marginalization are vulnerable to disasters, which exacerbate problems of poverty, indebtedness, and food insecurity. Many of the world's poorest are forced to live on unstable hillsides or in areas prone to drought or flooding.

Disaster planning must deal with the three prime unknowns, the "known unknowns," in Donald Rumsfeld's unfairly denigrated phrase: where and when such a disaster will happen; what unknown and unforeseen consequences will follow; and how to get governments and the public to take the whole thing seriously *before* it happens. After all, experts had warned that the Carolina barrier islands were no place for development; other experts had been warning for years that a direct strike on New Orleans's levees by a powerful hurricane would cause immense damage; yet others had been warning for decades that Kashmir was long overdue for a major earthquake. In none of these cases were they taken seriously. The "science" simply did not get through to policy makers.

It is not possible to build systems against all possible disasters. In certain cases, the solutions may be known but the costs of implementing them too high. Can you "move" New Orleans? Possibly. But Venice? And can you—should you even think about—"moving" Manhattan just because it would be horribly vulnerable to a Category 5 hurricane, or, in the longer term, to rising sea levels? And if you could contemplate such a thing, where would you put it? Where on Earth is risk free? On a smaller scale, cost-effective ways exist of minimizing the destructive effects of hurricanes and typhoons, among them harnessing the damping effect of wetlands, not building on fragile beach ecosystems, building cities farther inland away from storm surges, and implementing hurricane-proof building codes. It is even possible to build structures that are impervious to the 250-mile-an-hour vortexes of tornadoes, but the ratio of cost to potential damage means that we shouldn't even try. It would be ridiculous for everyone to live in a bunker against the very long chance that their house might find itself in a tornado's path—we just have to accept that in some circumstances, and in some scenarios, some fatalities are inevitable.

We have to keep in mind, also, what is ethically acceptable in a globalized world. Those luxurious bunkers deep beneath the Greenbrier resort in West Virginia, constructed to withstand a thermonuclear blast and to screen out the consequent radiation, were a refuge for precisely the politicians and scientists who made those blasts possible in the first place. We need to guard against any modern equivalent.

We need first to assess the risks, and then to manage them.

R isk management is hardly a new idea. Indeed, it goes back to the very earliest humans. Any hunting group that posted a lookout while other members were stalking game was deploying risk management techniques—they were guarding against being nastily surprised from the rear, either by rival hunters or predatory animals. In industrial Britain, miners used canaries in cages to warn them against the buildup of hazardous gases.

The capabilities—and potential destructiveness—of modern science have made these simple remedies obsolete. When planet-wide devastation is technically possible, the old protocol for risk management—try it once, if it turns out to be dangerous don't do it again—is no longer tenable. We need to find ways of warding off catastrophes caused by the development of modern technologies, whether nano-engineering, genetic engineering, or particle physics. We need to examine the ethics of conducting experiments for which there is no known outcome but which could have catastrophic results. The builders of the first thermo-nuclear devices believed nothing catastrophic would happen when they were detonated, and they were right—but they weren't really sure. How sure were the scientists at the Brookhaven National Laboratory that their heavy-ion collider would have only benign results? How do we govern this?

Risk management for natural disasters, as opposed to potential human-caused ones, can minimize their impacts by making informed decisions about where and how people should live, by sharing information among governments, and by investing in scientific research. For some calamities, such as hurricanes or volcanoes, predictive science is already well advanced. For others, like earthquakes, decades of research can suggest, more or less, where they may happen, but not when.

Even so, disasters will happen, and the international charity industry, weary as it is, will once again come forward to mitigate the aftermath.

It sometimes seems that the biggest threat to our well-being is not our lack of knowledge but our limited ability to actually make use of what we know, largely because of vested interests. An example is the suggestion from Aromar Revi, a New Delhi disaster mitigation specialist and consultant to the Indian government, who proposed a publicly accessible online database, something like Google Earth, that would map the risk landscape down to the neighborhood level, which would enable nations with shared risks to build better warning networks. But India, for one, has refused to participate. The Indian government can see the worth in a fully accessible tsunami early-warning network, but the sensors that network would need would also be able to detect Indian nuclear testing, which it finds unacceptable. National security is more important than the (now surely reduced?) risk of another tsunami that would kill thousands—this is a fairly typical government calculus, shared by political authorities worldwide. The fact that it is quite mad is beside the point.

Another suggestion has been to set up a global disaster relief insurance scheme, into which all at-risk nations would pay, thus transferring the risk to the financial industry instead of relying on the well-meaning but always fickle and short-term attention of international charities. Just such a scheme has been proposed by Reinhard Mechler, an economist at the International Institute for Applied Systems Analysis in Laxenburg, Austria. It, too, is unlikely to find fruition. It, too, would demand that scientists produce a risk-probability map, with the inevitable disagreements such a map would generate.[3]

Still, scientists in a number of disciplines are quietly working away at improving the planetary knowledge base. The study reported in Chapter 7, by Steven Ward and Erik Asphaug, on what would happen if an asteroid hit the sea sometime in the future, was done precisely for this purpose. The simulation was based on a real asteroid known to be on course for a close encounter with the Earth eight centuries from now. As Asphaug put it, "The way to deal with any natural hazard is to improve our knowledge base, so we can turn the kind of human fear that gets played on in the movies into something that we have a handle on."

Environmentalists have frequently argued for the invocation of something called the precautionary principle, which essentially says that governments and regulators should take steps to protect against future or potential harms, even if the causal chain is unclear, and even though we don't know that those harms will actually occur, or how serious they might be. Forbidding super-collider experiments on fundamental matter would fit into this category—invoking not only Donald Rumsfeld's known unknowns, but his unknown unknowns too. Drastic actions to reduce greenhouse gases are another case in point, if rather different in knowledge—an unknown known, if you like, shifting in many minds into a known known. The problem is, any action to combat these dangers would itself have uncertain consequences, and therefore the precautionary principle should be invoked for that action too, with consequent policy paralysis.[4]

A bout some of the specific natural perils facing our planet, we can do very little. About others, we can do a great deal. We need to learn to differentiate between these kinds of calamities, to learn to understand and to some degree to predict them. Then we can change public policy to minimize them and adjust our lives to deal with them and their aftermaths. How unprepared we are to do any of this was starkly revealed by the Asian tsunami of 2004 and above all by the bungling and recriminations that followed the failures to deal with hurricanes Ivan, Katrina, and Wilma in 2004 and 2005. That Katrina was so bungled is a bad sign, because hurricanes are perhaps the easiest of all natural calamities to anticipate and survive. The SARS outbreaks of recent years and the recent (and ongoing) scare over avian flu are other indications that we need better warning systems and emergency plans.

A steroids and comets are an interesting case, because they represent an example of very long odds (an impact is not very likely) with very severe consequences if a collision were to happen—at the very least, millions of casualties and, at worst, a civilization-ending catastrophe. This makes public policy difficult—how much time and effort, and how much money, to devote to surveys and precautionary measures? And once you've found a problem, what can be done? If anything?

The Planetary Defense Conference held in the United States in 2007 considered dozens of options, ranging from space tugboats to "nuking the bastards," a solution made popular by Bruce Willis and movies like *Deep Impact* and *Armageddon,* but generally rejected by scientists as too hit and miss—instead of destroying the target, bombs risk breaking it into multiple but still large and now radioactive pieces. (Also, which country was going to be encouraged to stockpile the weapons, obviously capable of space-borne deployment?)

Deflection seems to be the solution of choice. This could be done in a variety of inventive ways, including actually nudging the projectile out of the way with a spacecraft (hazardous to the crew); focusing mirrors on it to change its albedo and thus its eventual orbit; towing it off course with a giant lasso; or (the preferred solution of people like astronaut Edgar Lu) moving it off course by using only the minute gravity of a spacecraft as a "tow rope," the gravity tractor solution.

A former Apollo astronaut, Rusty Schweickart, is a member of a group calling itself the B612 Foundation (named after the asteroid home of Antoine de Saint-Exupéry's Little Prince), which has suggested a mission to send such a gravity tugboat to a nearby asteroid to test the scheme's feasibility. The European space agency also has a plan in place to test such a scheme. With a nice sense of irony they have named it Don Quixote, and its two rockets Hidalgo and Sancho.

All these schemes are feasible, it seems, given adequate warning time. Advance notice required would vary depending on the difficulty of the task. Maybe a decade. Maybe a century. Early identification is obviously critical.

NASA's Sentry and Safeguard efforts to catalog possible impact asteroids mostly focus on objects greater than a kilometer across. In 2006, the agency was given a year to prepare plans for a new survey that would catalog all Near Earth Objects at least 460 feet across; in March 2007 they delivered a paltry 27-page sketch with no detailed analysis, budget, or implementation plans; their attention had been focused, by presidential directive, on the rather sexier notion of getting humans to the moon (again) and to Mars. The report merely suggested continuing the current $41-million search program. Researchers in the field were scathing. Clark Chapman, of the Southwest Research Institute in Boulder, Colorado, had already estimated that tracking down asteroids as small as 460 feet would cost around $820 million for a survey using ground-based instruments, or $1 to $1.3 billion using a dedicated infrared telescope in an orbit near Venus, which could reach the survey goal three years faster and would be more sensitive. NASA, he said, showed no interest. "They [simply] didn't study the case of a multi-decade warning of a 325-foot object"—which would account for 95 percent of potential impacts and which could easily be diverted, given enough time, using a gravity tractor or other means of deflection.

"Nothing is perfectly safe in this world," Chapman acknowledged to the U.S. House of Representatives. "But if, ten years from now, we could say that we have reduced our worries by a factor of ten, that the chances of an asteroid striking are ten times less, because we have discovered and certified 1,800 of the 2,000 potentially dangerous asteroids as safe, then we could sleep a little easier at night. Moreover, if, by bad luck, there really *is* an asteroid headed our way, there might, after ten years of searching, be an excellent chance that we would have found it. And then, we could probably save ourselves."

The political problem is not trivial. The earlier a gravitational tractor is dispatched, the smaller and cheaper it would need to be. A longer wait means a more accurate gauge of the asteroid's trajectory and thus of the need to do anything at all, but the "doing" would be more expensive.

Nor is it clear who should make the choice to spend the hundreds of millions it would cost. The UN, working as *The Economist* puts it, at its usual glacial pace, seems an unlikely choice.

Of the Earth's inbuilt perils, volcanoes represent a case in which the hazard is easy enough to find (and even to some degree to predict) but impossible to stop.

There is no longer any need for a moderate eruption to kill anyone. The eruptive precursors are now well enough known and, at least in most cases, evacuation procedures could be implemented, at least for volcanoes whose eruptive and pyroclastic flows are the scale, of, say, Pinatubo in the Philippines. The tricky technical and political question is exactly when to move residents away from a possible blow. Too late, and major casualties are ensured. Too early, and people's lives will be disrupted for no apparent reason. A false alarm is the worst case, making subsequent alarms that much less effective.

With larger volcanoes, "resurgent calderas" such as Yellowstone, humans can do nothing but watch. If one of those were to erupt, it would cover an area far too large to evacuate and would have long-lasting effects on climate, with consequently devastating effects on human civilization. Perhaps our governors could save themselves in some equivalent of the Greenbrier Bunker in West Virginia—but save themselves for what, exactly? Still, a few precautions can be taken. For example, seeds, spores, and DNA markers are being stored in "banks" deep in the ice in Greenland and other secure places, which could then be used to reseed a devastated landscape when the time came. If there was anyone left to do the seeding.

Earthquakes are different. We can tell where they are likely, in a general way, but not precisely where, and we cannot yet tell when. Even so, earthquakes can happen anywhere on Earth; no part of the planetary surface is immune. (As though to reinforce this notion, the day before I wrote this passage late in 2007 a small earthquake of 3.2 magnitude happened in Bridgewater, Nova Scotia, not 20 miles from my house, a place that had never experienced a quake before.) Earthquakes kill many times more people than volcanoes do. But unlike the larger volcanoes, they are not civilization-threatening disasters, and that they do in fact kill as many as they do is to a large degree a matter of human carelessness.

As an illustration, compare the earthquakes that struck San Francisco in 1906 and Messina two years later. San Francisco had a population of 355,000, but only 700 died (no matter for the moment that the city was largely destroyed in subsequent fires). Messina had a population only half as large, some 150,000, but almost two-thirds of those were killed, more than 100,000 people—although Messina's earthquake was far weaker (7.5 to 8.25). The reason was very simple: San Francisco was built of wood, Messina of stone. Stone cannot stand horizontal shear; wood is resilient and bends.[5]

Our friends who had been in a beachfront hotel in Santa Monica during the Northridge earthquake survived, as did everyone else in the hotel—the structure, which was new, had been constructed on a series of Teflon "rollers," giving it a resilience it could not have had in any other way, and though they described the floor of the hotel corridor actually rippling in the quake, nothing ruptured. Even the seventh-floor swimming pool remained intact. Josh Jensen, the California winemaker whose winery is on the lip of the San Andreas fault, once saw his four-ton grape press "spinning like a top" in a quake, but the winery itself, anchored into a steep slope with horizontal concrete beams deep into the Earth, survived intact.

In short, buildings can be constructed to withstand all but the most severe of earthquakes.

In practical terms, it is impossible to retrofit a major building, and even more so a whole city. Tokyo, at the junction of three tectonic plates, is exceptionally vulnerable to earthquakes. Traditional Japanese buildings were made of wood and paper and, while easily set on fire, were at least earthquake proof; but Japan has followed the modern style of steel-and-glass high-rise buildings; most of the buildings would survive a quake, but the glass, were it to peel off, could cause numerous casualties. And in other parts of the world, such as the vulnerable Kashmiri region, construction standards are, as reported in Chapter 8, "among the scariest on Earth," with no chance that they will be improved. In that case, prediction and excess caution is all that could turn a disaster into a human catastrophe.

Prediction, though, is difficult. Exactly where and when a fault will suddenly give, producing a quake, is still not known. Precursor microquakes are thought to be a reliable predictor, but even then the timing is uncertain, the location only a guess, and the magnitude a mystery. Gases seeping from the Earth indicate that seismic activity is going on deep below the surface, but in most cases, nothing else happens, at least not for years and even decades.

For centuries people have noticed that certain animals seem to get restless before a quake, and science has studied the behavior of dogs, birds, and, on the theory that they are particularly sensitive to ground movement, snakes. A Japanese study found tantalizing evidence that catfish behavior changed in the runup to a quake, but the changes were small and inconsistent, and the research has been dropped. Other studies have tested chemical changes in groundwater, with the same result.

A potentially fruitful line of inquiry has been way off the ground, in the ionosphere. The ionosphere is a flexible, dynamic, and rather fragile system buffeted about by electromagnetic emissions, by variations in the Earth's magnetic field, and by the acoustic motion of the atmosphere itself, which means it is acutely sensitive to atmospheric changes. NASA has suggested that "there is persuasive evidence of an ionospheric precursor to large earthquakes" and that ionospheric changes could be used as an earthquake predictor. This is partly because acoustic waves are generated both before and after earthquakes, but also because it is thought that part of the runup to an earthquake is the generation of electromagnetic emissions, which have been detected in the ionosphere up to six days prior to a large quake. In other words, if we learn how, we might be able to use the ionosphere as an early-warning system, dramatically increasing the time people living in earthquake zones have to react. Even a day's warning could have made a massive difference, say, to the death toll of the Asian tsunami the day after Christmas 2004.

Tsunamis, because they are mostly caused by earthquakes, are equally difficult to predict. But of course unlike earthquakes, their effects are not instantaneous; a tsunami can travel at very high speeds, but the full effects, and the highest waves, usually appear an hour to six hours after the precipitating event, depending on location. As already reported, North America is well protected by tsunami sensors, and a global Earth observation system is being developed, which would give a globally comprehensive, sustained, and integrated watch on what goes on around the world; it is this system the Indian government decided to decline, on the grounds that it would catch them at nuclear testing. A monitoring system using older technology has existed in the Pacific since 1946. It has caused fifteen false alarms and five real warnings.

Reasonable precautions can be taken to survive all but the very largest tsunami. As reported, parts of Indonesia are using the loudspeakers in village mosques (used by the muezzin to call the faithful to prayer) as a warning network. Villages and housing developments should not be built on low-lying ground or on beaches. If possible, accessible high ground should be nearby. Natural barriers to tsunamis, such as mangrove swamps, barrier reefs, and lagoons, should be left intact.

These are the same precautions that will protect communities from hurricane-caused storm surges. Hurricanes can raise coastal waters by up to 23 feet, and on top of that you can have massive battering waves, all superimposed on a high tide. Since most hurricane damage is caused by storm surges and flooding, to be safe no housing development within miles of the coast should be located lower than 33 feet or more above high-tide level. Katrina, as already suggested, was a terrifyingly strong hurricane (though it was a Category 3 when it made landfall, downgraded from an awesome Category 5), but most of its damage was not wind-caused but water-caused, a consequence of poor planning and siting.

Most hurricanes are survivable, given proper planning and building codes; structures can be engineered to withstand very severe winds, and the forces high winds exert are well understood. A survey of Dade County, Florida, homeowners after Category 5 Hurricane Andrew had passed by found that many of them had taken refuge in a bathroom or closet in the middle of a house. Such spaces, however, provided little more than emotional cover in a storm that was known to have driven 2-by-4s through concrete walls. Much of Florida was considering adopting a code that would require new houses to have at least one room thought to be projectile-proof. Under consideration was an 8-foot square room sheathed in 2-by-4 studs covered with 4 inches of plywood.[6]

Over the years, there have been many attempts to control the ferocity of hurricanes, all so far futile; that none of them have worked never seems to discourage the optimists. The most recent—buttressed by supercomputer models and a little empirical observation—are notions to reduce the temperature differentials between the extremely cold convection currents at the top of a storm and the warm sea surface temperatures at its base, a differential that gives such storms their energy. The warming would be done by scattering planeloads of heat-absorbing black soot into the upper levels of the storm; the cooling, by seeding its lower reaches with salt crystals, which would condense water vapor and then evaporate, cooling the hot air. The researchers, including Ross Hoffman of Boston and Daniel Rosenfeld of Israel and Colorado, believe they could, at least, reduce a storm's intensity. This would be no small achievement: softening a storm from Category 5 to Category 2 would have a huge impact on the storm's damage toll. In litigious America, though, lawyers may represent a bigger threat than the storm. What happens if you reduce a storm's fury but end up diverting it to a place it would not otherwise have gone? Every property owner thus affected would sue. Worse, if you have the technology to divert a storm but don't, the property owners hit by the storm's natural track may sue anyway.[7]

Tornadoes are ill winds of a different kind. You cannot engineer an aboveground building to withstand tornado winds, or at least not one that anybody would want to live in. Troglodyte dwellings are probably immune, and shelters built completely below ground are probably immune too, but nothing else is (although studies have suggested that windowless concrete grain silos, which are round, have a good chance of withstanding a tornado). It is the pressure drop that causes the damage, not the inertial force of the wind itself. The atmospheric pressure can drop two hundred millibars within seconds; a house has no time to adjust and, quite literally, explodes outwards. Basement bathtubs are the refuge of choice, but where there is no basement you just have to trust that the tornado, whose path is after all quite narrow, hits someone else's house and not yours. No public policy but prediction can protect the public from tornadoes. Emergency relief after the fact is the only reasonable response.

Tornadoes are born of thunder cells, but their actual formation remains a mystery. It is known that where the preconditions exist, tornadoes can be formed within minutes. A network of Doppler radar units covers much of America's Tornado Alley, but the rupture in the warm-air cap that can produce tornadoes can be too small for the radar to easily see. As a consequence, sightings from the public and from amateur tornado chasers are taken seriously by professionals at the National Severe Storms Laboratory. With luck, warnings can come up to fifty minutes before a tornado strikes, but they can be issued as little as a dozen minutes before zero-point, perilously little time to take shelter, if, indeed, any shelter is to be found. Tornadoes seldom last for more than an hour.

Floods, like hurricanes, cannot be prevented, but the damage they do can be mitigated. Floods are caused by heavy rains, and heavy rains are simply a fact of life—they can happen anywhere. But rivers overflow mostly because we have re-engineered their flowpaths and have colonized their floodplains—the Rhine is a good example, and so is the Mississippi—and have built villages, towns, and subdivisions where even mildly abnormal weather would inevitably cause them grief. Sometimes this is a result of poverty. For millions of people in the undeveloped world, rivers are their source of both sanitation and drinking water and serve as a transportation route too, and they must perforce live on its banks. In the developed world, there is less excuse. But sometimes even there, local and regional authorities are trapped. A small town near where I live, Liverpool in Nova Scotia, is in such a bind. The town dates back to the mid-1700s; it was known as the port of the privateers in colonial times, when dozens of ships ventured from its safe harbor to prey on American shipping. In more recent years, two hydroelectric dams have been built upstream on the Mersey River, which flows through the town. Were those dams to burst in an extra-ordinary weather event (one nearly did in 2003, when more than 4 inches of rain fell in a few hours), the town would inevitably be flooded. But what to do? You can't move a whole town, and the hydro-electric power, in the modern world, is absolutely necessary; no one would contemplate decommissioning a producer of clean energy and substituting a coal-burning plant on the purely speculative grounds that some putative catastrophe might theoretically happen. . . . This is how public policy works. You don't take risks because you want to. You take them because it is easier, and less expensive, to do so and hope catas-trophe happens on someone else's watch.

Sometimes a simple policy change is all that is needed. After the British floods of 2007, blamed variously on climate change and government bungling, it became clear that up to 60 percent of the country was at least liable to occasional severe flooding—including vast numbers of new towns and subdivisions. The reason the housing industry built on such places, and the reason consumers so blithely bought the houses they built, is that under British insurance law gov-ernments, not insurance companies, are liable for damages caused by flooding. You could buy a house on a floodplain sure at least that were it to be flooded at some point, compensation would be forthcoming. This is beginning to change; some local authorities are beginning to balk and changing the rules accordingly. This will make insurance harder to come by, and therefore make housing more expensive. This, in turn, will deter builders from building in such places.

No place on Earth is more concerned with flooding than the well-named Low Countries; it's no secret that the people of the Netherlands, and to a lesser extent Belgium, have flirted with disaster for centuries. Rising sea levels are an obvious challenge.

In 2005, the Rotterdam Architecture Biennale held a major exposition called, simply, The Flood, which solicited technical solutions to the problems that will inevitably ensue. Some of the ideas presented included a floating soccer field, housing built on sponge-like synthetics capable of absorbing flood water, and entire floating subdivisions. About fifty floating homes have already been built on the banks of the Maas River, in a former RV parking lot. They are built on pontoons that project above the surface at low water, but that rise as the water does; a floating city for twelve thousand people is planned for the region near Amsterdam's Schiphol airport; the trickiest design problem, ironically, is providing clean drinking water.[8]

No one thinks this is a long-term solution. But no one is thinking of long-term solutions either. In the end, you can't keep out the rising sea. And you can't abandon a whole country. Until you have to.

P andemics can sometimes be contained, if not ended, by simple preventative measures—AIDS by safe sex and condom use, malaria by the use of cheap bed nets (a mass distribution of free mosquito nets in Kenya halved child deaths in less than a year). A little bit of money and a good dose of education would work wonders. In the rich world, drugs have turned AIDS from a certain killer into a chronic disease; a little bit of money won't do here—infusions of massive amounts are needed to do the same worldwide. Other diseases are trickier. This was graphically illustrated in May 2007, when a U.S. tuberculosis patient named Andrew Teacher set off an international uproar (and a dozen lawsuits) after defying medical advice and flying internationally on a wedding trip, putting hundreds of people directly and thousands indirectly at risk; he was home (via the Czech Republic and Canada) before anyone knew he was a problem. And if avian flu ever crossed the species barrier to become a stage 4 disease . . ."The strategy has to be looking at how to contain it in the animal world, because once you get it into the human world you're dealing with vaccines and antiretrovirals, which is a whole new realm," Nina Marano, a public health expert with the Centers for Disease Control and Prevention, told the Associated Press in 2006.[9]

"Containing it" in this context means several drastic measures— killing all infected birds and their whole flocks; sterilizing entire farms and sometimes regions of all possible birds; remaining vigilant for its spread through migratory birds, and pouncing when a dead bird is located; and monitoring the health of migratory flocks. The precautionary principle was already being applied: when in doubt, kill first and analyze afterwards. In 2006 and again in 2007, bio-labs and Big Pharma were gearing up for massive runs of production vaccines; medical and nursing personnel were first in line to receive them, and after them those concerned with public order.[10] Plans were being drawn up to ban international travel and to close national frontiers, with the awareness that none of this will stop the spread of the virus, only delay it some.

Authorities in many countries have contemplated the notion that they might even have to forbid inter-state or inter-provincial traffic, or even inter-city travel, to safeguard uninfected areas. This may prove impossible; modern economies are now locked into just-in-time distribution systems, so supermarkets and factories no longer keep inventories larger than necessary for more than a day or so—if the trucks and trains stop rolling, the economy stops too.

At the center of pandemic control efforts is, as it should be, the World Health Organization, based in Geneva. WHO has successfully headed off at least six disease outbreaks in recent years, all of which could have led to global pandemics. In June 2007 the agency's hand was strengthened by an international treaty, signed by all members of the UN, that obliges governments everywhere to report to WHO's director general, Margaret Chan, potential pandemics and outbreaks of infectious diseases "at once," which means within twenty-four hours.

A clever way of learning about pandemics and the way diseases spread is through the Internet gaming community, by introducing

infectious diseases into online games. This curious notion was suggested by Ran Balicer, of Ben-Gurion University of the Negev in
Israel, who professed himself inspired by a virtual plague called "corrupted blood" that swept through the online fantasy game World of
Warcraft (WoW) in 2005. The game's administrators had introduced
the virus as a deliberate complication. To Balicer's fascination, the
"corrupted blood" spread much faster than expected, partly because
household pets were made the vector. As Balicer put it in the March
2007 issue of the journal *Epidemiology*, scientists could work with
administrators of other games to release infectious agents of their
choice, controlling modes of transmission, symptoms, transmissibility,
and other factors to see how they spread and suggest ways they could
be controlled. He planned to talk to the administrators of the game
called Second Life, in which millions of people live surrogate lives of
their own designing, as a possible test site, "because it is much more
like the real world than WoW."[11]

Clearly, the battle is far from over. Bacteria are, and are likely to
remain, the planet's dominant species, and many of them are as agile
as the flu virus. They continue to probe what defenses we have
mounted—immunity, antibiotics, and the rest—and their evolution
doesn't stop. The Centers for Disease Control website has warning
pages for a plethora of diseases, including the newly worrying Lyme
disease and the West Nile virus (including a severe neurological
form), but also including plague, domestic arboviral encephalitides,
Chikungunya (a debilitating illness caused by infected mosquitoes),
Japanese encephalitis, tularemia, yellow fever, and dengue fever. This
last is, after malaria, the most widespread mosquito-borne viral disease affecting humans; an estimated 2.5 billion people live in areas
at risk for epidemic transmission. Each year, almost a million new
cases occur, with a mortality rate of somewhere around 5 percent.
It is carried by a different mosquito from the malaria carrier, *Aedes
aegypti*, which spread from Africa on cargo ships in the eighteenth
and nineteenth centuries, and by the early twentieth was endemic
in tropics everywhere. It proliferates easily in dense populations;
modern transportation systems seem purpose built for its expansion.
A virulent outbreak occurred in southeast Asia in 2006 and 2007,

spreading through Cambodia, Vietnam, Thailand, and even squeaky-clean Singapore, and turning up within weeks in the Philippines.

In 2006 the *New England Journal of Medicine* carried a report describing a case of multiple antibiotic resistant bubonic plague. The causative agent, our old friend *Y. pestis*, somehow acquired what is called a resistance plasmid from an unknown source.[12] And around the world, new strains of other infectious agents are turning up, the superbugs, including those causing tuberculosis and meningitis.

Then, as 2007 spilled over into 2008, a new and more dangerous form of Ebola emerged in Uganda. It was just as deadly as previous strains, but killed more slowly, thereby giving the disease more time to spread.

I f a pandemic were to break out, or some other calamity to occur, the authorities will tell you not to panic. This will be good, if rather ineffectual, advice. It will also come far too late.

But you don't want people to panic ahead of time either. This would not be good public policy. Who wants to live in a society in which fearfulness is the dominant tone? No citizens want to live their lives fearful of an errant bomb, a diseased bird, a mutating parasite, a hurtling meteor, a smoking mountain, or to live in a world where clouds are a source of anxiety instead of wonder, and birds a source of fear instead of joy. Good information is the best antidote. If you have good information, panic turns to planning, and proper planning implies that you can put quotidian worry out of your mind. Good public policy demands an informed citizenry, a cadre of specialists whose job it is to worry for you and to do what planning is necessary, and governors prepared to take what precautionary and then after-the-fact actions seem prudent. Sometimes all three of these seem in short supply. But really they're not. Citizens are becoming better informed. Scientific journals are publishing the necessary research, from hundreds of scientists, that will allow proper planning. And at times and in places, our leaders are beginning to listen.

SEVENTEEN

Making Things Better (ii): Undoing Human-Made Calamities

Risk comes from not knowing what you're doing.
—Warren Buffett

We know now that our species is making natural calamities worse and, in addition, is causing "natural" calamities of our own. Still, some things can be done to limit the damage and mitigate the risks. We can control our own actions, at least to a degree, and thereby make things better—we can "bring healing," in the cliché of the grief-counseling industry.

At least we *can*. It remains to be seen if we *will*.

We don't need a probability theorem to tell us when or how global warming and climate change will happen. We already know the answers to all the important questions. Even so, global warming is politically and psychologically the most difficult issue we face, partly because we are causing it ourselves, and partly because of the rhetorical mismatch between the apocalyptic warnings being issued every day and the cheerfully mundane remedies suggested (use less hot water, check your tires, hang your clothes on a washing line), which only serve to make the problem seem even more hopeless.

These mundane remedies will help—almost anything will help—but will not solve the problem. For that, we need larger and more urgent efforts, by science as a community, by governments, by corporations, and by concerted citizen action. As already suggested, nothing we do, not even on a large scale, will prevent the change from happening, unless we fix the population problem and the orientation of our growth-manic economy and make the switch from fossil fuel energy to something closer to sustainability. We are close to a political tipping point in attitudes toward global warming (a recognition that it is in fact a problem requiring action), but we are nowhere near a tipping point on these background issues. Because we are not, prudent policy demands that we put a good percentage of effort toward mitigating the effects of change (adapting to them), rather than toward prevention.

A Green (or Discussion) Paper set out by the European Union in July 2007 illustrates just how difficult this will be. The paper suggests that Europeans will have to make further and steeper cuts in greenhouse gas emissions, but also to begin to adapt to the climate change that is already happening. The Environment Commissioner, Stavros Dimas, put it this way: "People all over Europe will increasingly feel the threatening effects of climate change on their health, jobs and housing, and the most vulnerable members of society will be the hardest hit." Using water more carefully and caring for the elderly in heat waves were among the remedies suggested. Note, however, the hopeful future tense employed. The discussion paper then suggested a major conference with "all the stakeholders," a wonderfully bureaucratic formulation, and listed four areas for action: early action to develop adaptation strategies in areas where current knowledge is sufficient; integrating global adaptation needs into the EU's external relations and building a new alliance with partners around the world; filling knowledge gaps on adaptation through EU-level research and exchange of information; and setting up a European advisory group on adaptation to climate change to analyze coordinated strategies and actions.

The text repeated the now-familiar warnings: glaciers are melting and low-lying ski resorts are threatened with closure; southern Europe is drying out and may become too hot for summer holidays; winter storms are likely to increase in frequency and severity, particularly in Western Europe; coastal zones, low-lying deltas, and densely populated river plains could be affected by more frequent storms and floods.

Nowhere in any of this was there any recognition that global warming was, well, global in scope. Nowhere was there any mention of the population issue (partly, to be fair, because the European population has in fact stabilized and is growing only through immigration). And it never occurred to anyone to mention the economy at all, except to point out how "growth" would be adversely affected. That growth was the problem remains unacknowledged.[1]

One reason the population issue has been undervalued as a cause of global warming has to do with the nature of exponential growth (or "geometric" growth as Thomas Malthus called it). Many of the world's policy makers were born in the 1960s—when the world was a billion or two people emptier than it is now. They know, intellectually, that population "in the third world" is growing massively quickly; they may even have seen the evidence for themselves. And they know, also, that despite population stability Europe is a lot more crowded than it was when they were children (more crowded and much more polluted). Even so, the astonishing pace of exponential growth is easy to miss, and the sheer scale of change is seldom acknowledged. A useful way of looking at it is to use the analogy of algae growing in a pond that doubles in size every day. Say it entirely fills the pond by the thirtieth day—how much of the pond does it fill on the twenty-ninth? The answer, of course, is only half—exponential growth is apparently slow at first, then massively fast at the end. That's what is happening to population.

Nevertheless, the problem is solvable, and without draconian measures like forced contraception or the one-child-per-family decree of the Chinese government (a policy that is having its own unforeseen consequences, producing a generation of young men—sons are more desirable than daughters—who will never be able to marry, because there are not enough women). "Family programs," as they are carefully called, do work. In 1950, Sri Lanka had the same population as Afghanistan, but it implemented a set of fertility choices, and as a result it will have one-quarter of the Afghan population a century later. In 1970, Bangladesh had 5 million more people than Pakistan, but Bangladesh focused on making family planning available in culturally acceptable ways and Pakistan did not. As a result, by 2050 Pakistan will have 62 million more people than Bangladesh.

The education and empowerment of women is a key. Given the choice, women evidently want to be more than baby-factories. This has long been evident in the developed world—Europe, North America, Japan, and selected other countries. In the rest of the world, it is already a development cliché that if you educate the girls, the problem will, at least to a degree, become self-correcting.

I remember chatting to a storekeeper a dozen years ago in the Kenyan town of Narok about his family. He had ten children, "so far," five with one wife and then five with another.

Had the first wife died?

"No, she was used up. So I put her aside and got another."

Kenya at the turn of the millennium had 6.7 live births per woman, though it has dropped some since then, but the population is still growing at 3.3 percent, the world's fastest, according to the Kenya National Museum in Nairobi. Even if the economy grows at a respectable rate, Kenya is likely to become poorer, not richer.

The museum has a permanent display on the population problem, which it calls the country's most serious. Lining the walls are cases showing the various birth control methods and exhibits explaining the procreation process. Busloads of eager young schoolchildren, in their British-legacy school uniforms, are toured through the exhibits every day, bussed in from surrounding villages. The museum regards this education task as far more important than the history it displays in other exhibition rooms.

I told the curator of the population exhibit what I had been told in Johannesburg, South Africa's greatest metropolis. There, the black population had historically been growing at a much faster rate than the white; popular expectations were that it would continue to do so, and that population growth would continue to worry the politicians. But to the experts' surprise, the black birth rate had fallen dramatically, to a point where it is now just at replacement level. And that is without the plague called AIDS.

"Yes," the curator said, "this is because all the girls are going to school. Teach the girls, and the problem is halfway to being solved. It can happen here too." She looked out at the chattering, eager school-children. "Too bad," she said, "what they really want to see is Ahmed." (Ahmed is the skeleton of the giant elephant that became a national institution in the 1970s; President Jomo Kenyatta had even gone to the lengths of protecting him through a special presidential decree.)

She was wrong, though. The boys wanted to see Ahmed, yes. But the girls were clustered around the birth control cases. Their small black faces, round and eager, were rapt. Boys will be boys, but these girls would control the future.

Two other points, one of them positive news, the other somewhat more ambiguous.

Buried in the UN's 2007 report on population, the one that suggested a global population of 9.2 billion by 2050, was a sentence that predicted the rate of increase would drop, by that same date, to replacement levels. The population would peak at around 10 billion. That would be very good news.

On the other hand . . . Those Western economies that had already reached replacement levels or lower were beginning to see signs that families were once again getting larger. This was a consequence of a benign public policy aimed at getting more women into the work-force. Working women generally had fewer children, because the demands on their time and energies were so onerous. By 2005, however, so many governments had made it so much easier for women to both work and have families that the trend started to reverse.

Our economic system is a far less tractable problem. Who can possibly be against "sustainable growth"? Even though, as any physicist will tell you, growth in a closed system is, quite simply, impossible.

For most of human history, a steady state was the normal state. Only in the last 250 years has a growth economy been the dominant mode, and only in the last 75 or so has it become unquestionable dogma. What makes contemplating change so difficult is that the period of growth has coincided with—and almost certainly stimulated—many other changes too, most of them overwhelmingly positive. People everywhere, rich and poor alike, are living longer than they did, are living healthier lives, are less likely to fall prey to infectious diseases; fewer people starve, and almost everyone has access to more and better goods, and so on. I remember an economics teacher at the University of Cape Town once telling me: "For anyone who expresses dismay at modernity, I have a one-word answer: dentistry. Who wants to go back to chisel-and-mallet dentistry?"

It is not just that growth is the underlying rationale behind every economic decision. Even the most fundamental issues, such as property ownership, are indirectly dependent on growth. After all, a large percentage of real ownership is concentrated in a very few hands, but this curious fact is seldom questioned, and even less does it cause a fundamental rethinking of economic relationships. The safety valve we have given people at the bottom of the economic ladder is the notion that at least there is a way up. In a no-growth steady-state economy, this will not be possible; a steady state implies a cap on both ownership and income. It is hard to see the economic owners, who in turn control politics, willingly agreeing to this. (This is not a critique of "capitalism." Statist socialist and communist thinking is just as dependent on growth as any other system—remember Nikita Khrushchev's boast that the Soviet system would "bury" capitalists in short order? He was not talking quality of life, but the acquisition of more stuff. Those systems still demanded growth; the only difference was that the growth would be controlled by "collective ownership," which proved even more ruinous than the capitalism it sought to replace.)

Steady-state economics is not without its advocates. I have already mentioned Herman Daly, its intellectual leader and co-founder (with Bob Costanza) of the Society for Ecological Economics, but there are others, including Garret Hardin (*The Tragedy of the Commons*), Kenneth Townsend, and E.F. Schumaker. Costanza, in an important paper in *Science,* argued that continuing economic growth was a fantasy.[2] The appealing notion that economies could continue to grow while reducing energy consumption "must be put firmly to rest beside the equally appealing but impossible idea of perpetual motion," he wrote. (Costanza has also worked on "ecological pricing" and has argued that to include the true costs of gasoline's noxious effects on the environment, each quart should cost at least $2—and that's without taxes.) Still, orthodox economics is consistently hostile to the idea—as Daly wryly acknowledges, economists have developed a single-word description of the consequences of a steady-state economy: stagnation.

One of the advocates for a steady state was the classical economist John Stuart Mill, who in his *Principles of Political Economy* put it this way:

> I cannot . . . regard the stationary state of capital and wealth with the unaffected aversion so generally manifested towards it by political economists of the old school. I am inclined to believe that it would be, on the whole, a considerable improvement on our present condition. . . . It is scarcely necessary to remark that a stationary condition and capital and population implies no stationary state of human improvement. There would be as much scope as ever for all kinds of mental culture, and moral and social progress; as much room for improving the art of living, and much more likelihood of it being improved, when minds cease to be engrossed by the art of getting on.[3]

As Daly points out, "A situation of non-growth can come about in two ways: as the failure of a growth economy to grow, or as the success of conscious policies aiming at a steady-state economy [SSE]. No one denies that when a growth economy fails to grow, the result is unemployment and suffering. The main reason to advocate a SSE is precisely to avoid the suffering of a failed growth economy because we know that, sooner or later, the growth economy will no longer be able to continue growing." Of course, recently the global warming debate has shifted this formulation from the purely theoretical to the alarmingly practical. Daly himself used the analogy of an airplane.

The fact that airplanes fall to the ground if they stop in the air merely reflects the fact that airplanes are designed for forward motion. It does not constitute a proof that helicopters cannot remain stationary.[4]

What *is* a steady-state economy? However defined, it will include all of the following: a constant human population; a constant stock of artifacts and manufactured goods (old ones replaced only as they wear out); a standard of living sufficient for a good life and sustainable for the long-term future; and a rate of "throughput"—that is, consumption of goods, by which the two stocks (human and goods) are maintained at the lowest feasible level. It does *not* mean no development—no improved technology, art, product mix, creativity, or wisdom. One way to encourage "constant throughput" and to discourage growth is to tax resource extraction with draconian severity, to the extent of generating most national income from this source, and at the same time to drastically lower income taxes. As Daly himself points out, this would allow optimists to use the higher price generated for resources to finance incentives for technological advances, while pessimists would be protected from their worst fears. A corollary is that non-renewable resources should be depleted only at a rate equal to the rate of creation of renewable resources.

But you can see why it is not likely to happen. First of all, most of the world's people, and therefore most of the world's countries, are poor, and for them "more" really does still mean "better" and will continue to do so until a threshold of development is met—say an arbitrary income of maybe $15,000 a year. The chances of this happening soon are, pretty obviously, vanishingly small. And even in the developed world, where increasing numbers of surveys show that more stuff makes us neither materially better off nor in any way happier, the prospects are still remote. To get to a steady-state economy involves rethinking the way our entire economy and society is organized. It implies some sort of mechanism to regulate resource depletion, either by taxes or quotas. It also implies some sort of governing institution that regulates minimum and maximum limits on income and on the accumulation of wealth. In addition, it implies zero population growth and therefore some way of supervising population levels (perhaps transferable birth licenses, suggested by an economist named Kenneth Boulding in 1964).

"Unlikely" therefore seems something of an understatement.

To illustrate how difficult it would be, consider one small case: population and economic growth in the face of increasing droughts and aridity in the American southwest. A report in the journal *Science* put it this way: "A broad consensus of climate models is that southwestern America will dry considerably through the 21st century and that this transition is already under way. The levels of aridity of the recent multi-year drought or the Dust Bowl and the 1950s droughts will become the norm within a time frame of years or decades."[5] So there will be, inevitably, increasingly severe water shortages in the region. In this context, the real question, asked by climate scientist Gregg Garfin at the University of Arizona, is this: whether governments should be pursuing new sources of fresh water (new diversions from afar, desalination factories, etc.) or considering limits to growth. But there is, and can be, no public or political debate on this issue—how can you stop Los Angeles growing?

Because it will not, doing something to mitigate pollution and greenhouse gases takes on ever more urgency.

The prime need is in the energy sector, because it is our use of fossil fuels that precipitated the greenhouse crisis in the first place. And here we can do a great deal. Even George W. Bush in 2006 called for ending "our addiction to oil," although to placate his own political base he carefully inserted the word "foreign" before the word "oil"; it amounts to the same thing, since the United States hasn't nearly enough for its own use. Then, in 2007, he summoned a sixteen-nation conference he called a "climate change summit." The same month the U.S. House slapped a $16-billion tax on the oil industry to pay for renewable energy and conservation incentives.

Greenhouse gases can be reduced in three ways: by using less energy through conservation and increased efficiencies; by finding alternative sources of energy that don't produce greenhouse gases; by doing business as usual but using geo-engineering to remove the carbon dioxide and methane from the air. All three areas have produced a flood of ideas from scientists, entrepreneurs, and ecologists—a review of the literature can give a wonderfully positive perspective of human creativity. All that's really needed now is a sense of urgency from our political leaders.

1. Conservation and increased efficiencies

Here are just a few of the many ideas that were circulating in 2007:

The efficiency of the national electricity grid (that is, the difference between energy produced and energy actually used by consumers) is only around 55 percent. "Smart metering," combined with a less centralized generating system, could capture energy in the 70 to 80 percent range, a massive saving. This less centralized grid would produce electricity with small regional turbines, probably driven mostly by natural gas.[6]

PHEVs, or plug-in hybrid electric vehicles, if produced on a large enough scale, could dramatically increase efficiencies. For one thing, they could be charged at off-peak hours and, because they are plugged directly into the grid, they could themselves serve as a massive highly distributed battery for the grid itself. Millions of such vehicles could be charged without building any new power stations at all, since existing plants seldom run at full capacity, except in heat waves and severe cold, and even then there are off-peak hours. A report by the Pacific Northwest National Laboratory in the United States suggests that existing grid capacity could power 217 million light-duty vehicles, three-quarters of all American cars. Switching to PHEVs could save 6.5 million barrels of oil a day, half the U.S. imports. Such a change would mean running existing plants at a more constant high rate, but because massive generators are more efficient than car engines, the net balance would be a 27 percent reduction in U.S. greenhouse gas emissions. The bigger batteries and more expensive vehicles could be financed by either a carbon tax or paid for by the savings on oil expenditures. James Woolsey, a former CIA director whose Set America Free foundation is dedicated to weaning the country off oil, admits that the owners of such cars "would have to make an investment in infrastructure: an extension cord."[7]

Over 70 percent of total energy consumed is used either for transport or in buildings, for heating and cooling. Numerous ways exist for increasing building efficiencies, most of them easy to implement and use. They include radically increased insulation standards, the use of geothermal heating and cooling, and, a really radical one, buildings with windows that actually open to let in breezes as opposed to mechanical cooling (a Canadian study found that schools whose windows opened had fewer sick days per child and teacher than mechanically ventilated buildings, another advantage). In-fill building and increased city densities are an easy and painless way of increasing efficiency—less transport use, buildings easier to heat and cool, more pedestrian activity, and many other advantages—the Paris model, instead of the Dallas model.

A modern cargo ship launched in 2007, the *Beluga SkySails*, carried a kite-like sail that was said to reduce fuel bills by up to 15 percent, and can be deployed sailing very close to the wind.

Green roofs—planting vegetation on flat or even sloping roofs—radically reduces the need for heating, and if used on a large enough scale can change a city's microclimate. A study in Minneapolis found that increasing a city's parks space (a roof can be so defined) by a mere 10 percent can change the ambient temperature in the urban area by a full degree. Other studies have spun this out globally, with the same conclusion—a 10 percent increase in green space could make the planet a degree cooler. Green roofs have the added benefit of reducing water runoff and slowing storm water drainage, thus reducing the need for expensive separation of storm and sanitary sewers, a considerable saving.

Sometimes large changes can be made with simple, no-cost policy decisions. A remarkable example comes from the African country of Niger, one of the poorest countries on Earth. Better conservation, brought about by one small change in government policy, and, yes, a small increase in rainfall, have created nearly seven million acres of new tree-covered land—satellite images show that Niger is considerably greener than it was thirty years ago, something achieved without large-scale plantings or massive foreign aid. What happened was simple. From colonial times all forests, and even individual trees, had been regarded as government property, which gave farmers little incentive to protect them. Indeed, most were cut down for fuel. So Niger's government turned ownership over to the peasant smallholders, who could now make a little money by selling branches, pods, fruit, and even bark. Because these sales are more lucrative over time than chopping down the trees, the trees are preserved and their growth encouraged. The ecological benefits are obvious: the trees fix the soil in place, preventing wind erosion, and they hold water. A few million acres is not enough to change the regional climate, but three times that amount might be.[8]

2. Alternative fuels

Most existing alternative fuels produce energy that is much more expensive than oil and need massive subsidies. Or so says the oil industry. But they are wrong on two counts. One is the implication that Big Oil exists without subsidy, which is a grotesque fabrication. The truth is that the real cost of production is shared by the taxpayers, while the profits go to the corporation. The biggest of these real costs is the expense of cleaning up what the use of the product has wrought—that is, pollution, smog, and greenhouse gases, with all the ills to human health and planetary well-being that they impose.

There is a whole basket of technologies that can, and in time will, replace oil. They all have their costs, and yes, they need subsidizing. Many of them are still small and, in some cases (wind, for example, and solar) dependent on a variable resource. Others are still under development—tidal power (for instance, in the Bay of Fundy, where the tides can reach 52 feet) and using the power of ocean currents are examples. Still other technologies are already substantial—ethanol and bio-fuels, for instance, though these still use more energy in their production than they produce. And a few are fully developed and massive in scale and could serve as a short-term replacement for oil. Nuclear power is the most obvious example.

Two other possible solutions to the energy problem and its associated greenhouse gases are held out as long-term prospects. Neither is yet feasible. The first is a hydrogen economy, with the hydrogen itself produced either by splitting water into its constituent molecules, leaving oxygen as a "waste product," or by generating electricity, which is then transformed into hydrogen by fuel cells. The other is nuclear fusion, the energy that powers the sun—and a hydrogen bomb. Fusion, were we able to control it, would leave no radioactive waste, unlike fission reactors. Late in 2007, scientists were tinkering with the notion that fusion could be produced by powerful and highly focused lasers.

A hydrogen economy has many attractions. Among them are the element's very abundance and its non–polluting aspects. Its disadvantages are the lack of a hydrogen infrastructure—oil, whatever its demerits, has over the past century built up a massively decentralized distribution network—and, as yet, no efficient way of producing it that doesn't itself use considerable amounts of energy—hydrogen is not easily detached from the oxygen in water. One suggestion is to use massive solar arrays in desert regions to generate electricity that can then power fuel cells to produce hydrogen. If long-distance DC power lines are used instead of the current AC grid, voltage losses would be minimized, and such arrays would become feasible. Another idea is to use fuel cells as a distributed grid. One of the great advantages of fuel cells is that they are reversible. They can be used to make electricity efficiently by burning hydrogen but can also be a source of hydrogen when supplied with electricity. Just as with PHEVs, the total number of cars on the road has a generating capacity many times greater than existing power stations; fuel-cell–powered cars can thus be the fuel store and generators of the national grid.

In the stark countryside of Andalusia, near the ancient Moorish city of Seville, is a slender and elegant white tower, forty stories tall, bathed in daytime in a dazzling halo of white light. It is a slightly surreal sight there in the parched Spanish countryside; the brilliant rays that light up the tower seem at first glance to emanate from the tower itself, as though it were some alien object touched down on the Earth. But instead the light comes from an array of more than six hundred finely polished steel mirrors filling several acres around the tower, focusing the sun's rays on the top third of the tower itself, where the light and the heat are intensely concentrated, forty to sixty times the heat of the sunshine, easily hot enough to produce steam to drive a turbine that then produces electricity. This is the power station of the future. Or if not that, one of the power stations of the future.

The technology is called concentrating solar power (CSP), and it differs from conventional solar cells, which convert solar rays to electricity through chemical processes. CSP doesn't use chemistry, just sheer heat. The Seville plant produces eleven megawatts of power, enough to power the city of Seville itself. The tower, so far the only one of its kind, is owned by the Spanish energy giant Abengoa, better known for its bio-fuels production. It produces not one gram of greenhouse gases.

Other CSP technologies do exist, though they generally take up more room and are not nearly as elegant. The most common is called trough solar systems, which use parabolic curved, trough-shaped reflectors to focus the sun's energy on a receiver pipe running through the reflector's focal point. The energy heats a transfer fluid, usually oil, which is then used to generate steam. The fluid is also routed through a thermal storage system that allows the plant to keep operating after the sun has set. Usually, this is a two-tank system, in which the heat transfer fluid passes through a heat exchanger, where it gives up a portion of its heat to a nitrate salt solution.

The capacity of plants under construction is not negligible. In 1994 the U.S. Western Governors Association put up funding to develop 1,000 megawatts of CSP power by 2010; a 64-megawatt plant is already operating near Boulder City, Nevada. Two equally large plants are being built near Granada, in Spain, and two more near Seville; in the Negev Desert in Israel a 150-megawatt system that is to be expanded to 500 megawatts is being developed by a company named Solel, which expects to spend a billion dollars in the construction. In 2007, Jordan was proposing an ambitious scheme to supply Europe with power from a vast network of solar generators located in the Sahara and Middle Eastern deserts.

The cost of electricity produced currently runs about 10 to 12 cents per kilowatt hour, more expensive than but competitive with conventional fossil fuel generation, and of course invulnerable to fuel price fluctuations The aim is to reduce this to somewhere around 4 cents by 2020.

Of course, the plants are expensive to build. But once built they never have to pay for fuel.

Alas, they don't work when it drizzles.

Conventional solar-cell technology is better for small-scale distributed power generation. Until recently, its efficiencies have been poor. But the U.S. Department of Energy reported in 2007 that solar cells have been developed with an energy conversion efficiency of 40.7 percent, with 40 percent the breakthrough rate for commercial use. The new cells, a co-venture between the Energy Department and Spectrolab, a subsidiary of the aircraft maker Boeing, are multi-junction cells, individual cells made up in a thin sandwich of different layers, each capturing a different part of the sun's spectrum. These cells may permit energy production at somewhere around 8 to 10 cents per kilowatt hour, still more expensive than the grid, but not by much.[9]

They don't work when it rains either. But they do work, sort of, in the fog—the Canadian government's weather station on Sable Island in the Atlantic, one of the foggiest places on Earth, produces all its hot water from solar panels, and so far the staff haven't had to forgo their showers.

Of all the alternative technologies, wind turbines are the most advanced. The advantages are the same as for photovoltaics or for concentrating solar energy—no more being beholden to OPEC, no more dependence on a depleting resource, no more carbon-based pollutants. The new "windmills" are a world away from the old ones that used to dot the European countryside and the American prairies. If you stand underneath one of them, it looms overhead, gigantic and otherworldly, as massive and unexpected in the countryside as an office tower. Indeed, these things are the size of an office tower—a single modern turbine is the size of a forty-story building, and a small wind farm of maybe two dozen turbines can easily generate thirty or more megawatts of power.

It is a simple technology, though it demands sophisticated engineering to keep it working reliably, in winds that vary from light breezes to hurricane force gales. In principle, turbines are merely the

opposite of a fan. Instead of using electricity to make wind, wind turbines use wind to make electricity. The wind turns the blades, which
spin a shaft, which connects to a generator and makes electricity. So
far so simple. What makes it effective is the force that wind generates.
The force exerted by the wind varies by the square of its wind
speed—that is, the force exerted by a 12 mile wind is four times the
force exerted by a 6 mile wind. It's not just wind that does this—the
ratio applies to all kinetic motion. If you double the speed of a traveling car, say, it will take four times the power to bring it down to a
standstill, in accordance with Newton's second law of motion. But the
wind's energy as it applies to windmills is greater still—it increases
with the cube (the third power) of its velocity, not the square. That is,
if the wind speed is twice as fast, it contains eight times as much
energy ($2 \times 2 \times 2$). This apparently puzzling fact is explained by the
fact that the wind passes through the turbine, so that if the speed of
the wind doubles, twice as many slices of wind pass through the turbine each second, and each slice contains the usual four times as much
energy, yielding up the eight-times result.

The speed of the winds, then, is critical and has great practical
implications. Take an average wind speed for a day of an arbitrary
11 miles an hour. Say the wind in London blows all day at that
speed, 11 miles an hour, but the winds in Paris blow 8 miles an hour
half the day and 14 the other half. The averages are the same, but in
practice Paris would get as much power from half a day as London
did the whole day. The shorter but faster winds add enormously
more power.[10]

As an industry, wind energy didn't really exist 20 years ago, but
there was a five-fold increase (487 percent) in wind-delivered electricity between 1995 and 2001, and wind power is now growing by
30 percent a year. The numbers are no longer trivial. The 15,200 new
megawatts of installed power in 2006 alone is enough to offset the
carbon dioxide emissions of 23 average U.S. coal-burning generating
plants—or somewhere around 8 million cars.

By Lester Brown's calculations at his Worldwatch Institute, by 2004
enough power was already being extracted from the wind to meet the
needs of 23 million people, or the combined populations of the

Scandinavian countries. Denmark gets more than a quarter of its energy from wind, a target only dreamed of in North America. The Spanish province of Navarre in 2007 did even better, generating almost 60 percent of its energy from renewable sources, almost all of it from wind, and planned to increase the number to more than 75 percent by 2010.

One of the downsides of wind power is that the wind doesn't blow all the time. But there is a move afoot to construct a Europe-wide grid of wind generators, which would get around that difficulty—wind is always blowing, somewhere.

Curiously, the actual deployment of wind technology has drawn fierce opposition, mostly from people who don't want the damn things cluttering up their view. Since they are sited where the wind is frequent, and since the winds are constant in some beautiful seaside locations, a good deal of this opposition has been from wealthy vacation home owners. But many Greens oppose them too, or at least oppose their presence in otherwise pristine, though expensively maintained, nature. Much of this has to do with aesthetics and not economics. To some, they are simply ugly, though presumably not as ugly as a coal-fired generating plant. But then coal plants are seldom, if ever, sited where they bother the high and the mighty.

The Dutch, ever practical, are solving this awkward dilemma by locating the wind farms so far offshore that they are barely visible on the horizon. Construction started in 2007 on just such a wind farm, 14 miles off the Dutch coast, and five to ten more were planned.

Bio-fuels—produced from a wide range of "biomass," including cow dung, soybeans, sugar cane, palm oil, sewage, and food scraps—are already a major industry. That more American cars don't run on Brazilian ethanol from sugar cane has nothing to do with Brazilian capacity, but only with pork-barrel subsidies for American farmers and regressive tariffs. More and more corn farmers are turning from food crops to fuel as their main source of income; research is under way to find, or "design," microbes that can improve the efficiency of fuels; more research is aimed at catalysts that might enable bio-fuels to be produced from ordinary wood, especially slash left over from logging operations, which would make this a truly massive industry.

Many disadvantages have to be overcome. More fuel is still consumed in planting and harvesting crops for bio-fuels than is produced; when the full life cycle of agro-fuels is considered—from land clearing to automotive consumption—the moderate emission savings are undone by far greater emissions from deforestation, burning, peat drainage, cultivation, and soil carbon losses. The ecological lobby group Greenpeace, which you would think would be in favor of alternative fuels, states flatly that "if even five percent of biofuels are sourced from wiping out existing ancient forests, you've lost *all* your carbon gain."

Still, the notion of using wood from trees, especially fast-growing trees like alder and poplar, is gaining ground. Scientists are aiming to genetically engineer trees with less lignin and cellulose, in order to make them break down more easily to liberate the sugars that make the fuel. Trees, of course, are perennial and ubiquitous, more versatile than food crops. And there are lots of them.

Other companies, like for example Codexis of California, are using synthetic biology to make fuels much more efficient than ethanol. Another firm, LS9, was looking to create "biodiesel," also using synthetic biology to create fatty acids that could easily be turned to fuel.

All these technologies—solar, bio-fuels, wind, tidal power, and the rest—have their own costs and their own inefficiencies of scale. In one notorious example, a study found that damming and flooding the entire Canadian province of Ontario would not yield as much power as the province's twenty-five nuclear power plants. To meet the U.S. energy demand with wind power would mean windmills covering the states of Texas and Louisiana. Almost 5 acres of land would be needed to produce bio-fuel for one car). "These technologies would require a massive industrial transformation of the landscape," says Jesse Ausubel, director of the environmental program at Rockefeller University in New York, from whose report these examples are drawn.[11] "The bio-fuels business is madness," he added.

Which brings us to the power of the atom.

A few minutes after 1 A.M. on April 26, 1986, after a series of grotesque accidents and management failures, one of the reactors at the Chernobyl Nuclear Power Plant near Pripyat, in Ukraine, went into meltdown and exploded, sending a deadly plume of radioactive contaminants high into the air. It was the worst accident, by far, in the not-very-old nuclear industry.

A few years after Chernobyl exploded, I was on a passenger vessel bound down the Volga River from Moscow to Nizhni Novgorod; and at the little town of Dubna, not far from Moscow, the vessel was boarded by a party of young people. I noticed them that evening, drinking silently in a corner of the lounge. Many of them were beefy young men. As the evening progressed, they got noisier, in the way groups in bars do, and occasionally women joined them, wives and girlfriends. The men drank many toasts, but clinked glasses only half-heartedly with the women; they grew emotional and touched each other in the way men do when they want to express emotion but don't know how. They'd rub each other's shoulders or pat each other's backs. I finally asked someone who they were.

"Oh," he said, looking over at them affectionately, in a proprietary way. "They're firefighters from Chernobyl."

The ship was not full, and its owners, as a gesture, had padded the lists with Chernobyl veterans. Some of the men on board had been in the first wave, the village firemen sent in a few minutes after the explosion to put out the fire in the containment building; one young man, the one in the corner with brooding eyes, had been the third person in the building, forty minutes after the explosion.

"And he survived?" I asked, incredulous.

"There he is, Sergei, ask him."

Later, I did, and Sergei gave me a small pennant as a souvenir. On the triangular silk are printed pictures of six of his colleagues who died a few days after the event, and the words "For Their Heroism and Sacrifice." Later, I spoke to the doctor who was traveling with them; she told me more about what they'd done, and the local bureaucrats' despicable attempt to blame the disaster on "worker neglect" (instead, it had become clear later, the engineers in charge had conducted unauthorized and unsafe experiments), and how they'd put together a Chernobyl victims group to press for more and better medical care—for more long-term care, for better chemo-therapy treatments—anything.

"They look fine now," I said, "they still have their hair, they look tanned and more or less fit."

"They are young, and strong," she said, "but they'll all die of cancer. They know that."

She told me she had been present when the authorities were forced to exhume the bodies of some of their colleagues. They had to be re-buried in lead-lined coffins: radiation was seeping into the ground, a deadly frost.

The next morning I was up early so I could see the vessel leave the Moscow-Volga Canal and enter the river proper. It was 4:30, the summer sun already rising, the early rays catching the outstretched arm of the statue of Lenin placed watchfully at the canal's terminus, and I heard a melancholy sound coming from the bow of the ship, a song of lament and sorrow. I crept forward to see who it was. One of the Chernobyl wives, bundled up against the morning freshness, her blond hair pulled tight into a bun, her eyes squeezed shut, was keening into the wind.

> Our way lies through the steppe, through
>   sadness without bound
> Your sadness, Rus, your tears,
> The haze of light, the haze of the beyond
> I do not fear
> The sunset and the heart astream with
>   blood
> No calm, not anywhere
> Weep heart! Weep loud! Past field, past
>   wood
> Gallops the steppeland mare ...

This was a poem by Alexandr Blok, she told me later. But the music was her own. To sing softly of the sunset in the Volga's early light was inexpressibly sad, and I left her there to her solitude, singing of her pain.

So what, in 2007, is an environmentalist with impeccable credentials like James Lovelock, he of the Gaia theory, doing calling for
a crash program to build ever more nuclear power stations, and build
them quickly? And why is David Suzuki, the nearest thing the
Canadian environmental movement has to a saint, refusing to con-
demn nuclear power out of hand? Why is *The Guardian*, that reliably
lefty and green-tinged newspaper, publishing pieces suggesting that
"atomic power is crucial in the fight against global warming"?

Lovelock makes three points: Look at the safety record, he says.
Look at the residual waste problem. Then look at the alternatives. And
you'll come to the same conclusion.

There have been industrial accidents in the nuclear industry,
although not many. In 1956 a military reactor at Windscale in the
United Kingdom caught fire and released what the secretive British
would later call "significant amounts" of radioactivity into the atmosphere. No one was told about the accident, and the British issued no
warnings. Nevertheless, epidemiological studies afterwards could find
no unusual deaths and no increases in cancer rates after the accident.
The accident seemed to have no real consequences at all. In 1962,
both the Russians and the Americans conducted atmospheric nuclear
tests that released into the atmosphere radiation equivalent to two
Chernobyls every week for a whole year. It was soon possible to find
trace amounts of a strontium isotope in the bones of pretty well
everyone in the world—and yet human lifespans across the globe have
continued to increase. Three Mile Island, near Harrisburg in
Pennsylvania, remains the only nuclear accident to have happened in
the Americas. The operators of the plant made a series of six grievous
errors within fifteen minutes, culminating in the loss of reactor
coolant and a threatened meltdown. It didn't kill a single person.

But Chernobyl? Surely it doesn't get any worse than Chernobyl?
In the months and years after the accident, media reports suggested
thousands, even hundreds of thousands, of Ukrainians, Russians,
Belorussians, and eastern Europeans were dying or at risk. Half a
million deaths was a figure that routinely appeared in press reports.
Then, twenty years after the event, in 2005, the World Health
Organization issued a careful assessment of damage: fewer than fifty

deaths had occurred to date, and almost all of those were rescue workers like those I had met on the Volga steamer.

This report, *Chernobyl's Legacy: Health, Environmental and Socio-
Economic Impacts*, was not a casual affair. It was a three-volume, six-
hundred-page report, incorporating the work of hundreds of scientists,
economists, and health experts from eight UN specialized agencies,
including the International Atomic Energy Agency (IAEA), the World
Health Organization (WHO), the United Nations Development
Program (UNDP), the Food and Agriculture Organization (FAO), the
United Nations Environment Program (UNEP), the United Nations
Office for the Coordination of Humanitarian Affairs (UN-OCHA),
the United Nations Scientific Committee on the Effects of Atomic
Radiation (UNSCEAR), and the World Bank, as well as the govern-
ments of Belarus, the Russian Federation, and Ukraine.

A common press headline said after its release, "UN suggests up to
400,000 people could die from Chernobyl."

Indeed, the report suggested that "the epidemical risk" of exposing
the population of Ukraine and Europe to 10 milisieverts of radiation
(about 100 chest X-rays) could reach 400,000. The same statistical
analysis said that the life expectancy of most of those would be
reduced by less than a week, and many by just a few hours—far less
than the damage caused by second-hand smoke. A high rate of thyroid
cancer was detected in children in the immediate neighborhood of
Chernobyl—but 99 percent of them survived. The biggest problem to
date has been the mental health of people in the surrounding area,
who have suffered from a high incidence of depression and anxiety.

In 2007 wealthy Russians were buying condos in the neighborhood, partly because it was a trendy thing to do, but partly because
the region, from which many humans had been removed, had
become a haven for wildlife of all sorts.

Chernobyl did have one disastrous result: it paralyzed the nuclear
industry and turned public opinion against nuclear power. In effect, it
made global warming worse by cementing our reliance on coal-fired
generators.

In the middle of July 2007 the nuclear industry—and the world—
was once again jolted when a major earthquake (magnitude 6.8)

occurred on the west coast of Honshu Island, badly damaging the world's largest nuclear power plant located, perversely, on top of a major fault line. Depressingly, the operators of the plant did what everyone seems to do when confronted with bad news—they denied it, underplayed it, and tried to cover up their errors. In the aftermath, it seems that the seven reactors successfully shut down, but fires were set off in several transformers and cracks appeared in a dozen places; the Japanese government shut the place down "indefinitely" until a major survey could be undertaken. The Tokyo Electric Power Co. at first said that only 1.6 quarts of radioactive water had leaked into the sea; later it admitted that they had not been candid and the figure was greater by several orders of magnitude. Even so, even in the short term, the consequences are likely to be minor. There were many alarums about "disaster averted," but at least it *was* averted.

So is nuclear power safe? Clearly not. But it is safer than, say, hydro-electric power or the coal industry (there have been 31 deaths in the nuclear industry per terawatt year, compared to 4,000 in hydroelectric schemes, and 56,400 in the coal industry).[12] Coal-fired plants routinely cause thousands of deaths every year, mostly from breathing coal dust.

What about nuclear waste, which notoriously lasts for millennia? Here, too, Lovelock is sanguine. "The waste generated by fossil fuels is a mountain a mile high and 12 in circumference. The same quantity of energy produced from fission would generate two million times less waste, and would occupy a 16 meter cube. The CO_2 [produced by the fossil fuel industry] is invisible but so deadly that [in time] it will kill nearly everyone. The nuclear waste poses no danger. . . . I have offered to accept all the high-level waste from a nuclear power station for a year for deposit on my small plot of land. It would occupy about a meter and fit safely in a concrete pit, and I would use the heat from its decaying radioactive elements to heat my home. It would be no danger to me, my family or to wildlife." Gaia, the Earth itself, doesn't at all mind nuclear waste, he says. It is much less destructive than, say, farming, or even pets. "The natural world doesn't mind nuclear waste. One of the striking things about nuclear waste depositories, either around Chernobyl, or in the bomb test sites of the Pacific or near the Savannah River, is the richness of their wildlife. To Gaia, nuclear waste is the perfect guardian against [the worst polluter of all] man."[13]

In this regard it is interesting that in July 2007 the U.S. government decommissioned the Rocky Flats nuclear weapons production area near Denver, Colorado (most of the trigger mechanisms for the U.S. nuclear arsenal were made there), and turned it into a wildlife refuge. It had been closed to the public for more than fifty years. The site had been contaminated with a witch's brew of compounds, including plutonium, uranium, beryllium, and several hazardous chemicals.

"Here we are," in the words of physics professor Jim Al-Khalili, of the University of Surrey, "trying to figure out how to avert the disaster of climate change now, and yet the long-term problem of nuclear waste still worries us. . . . I find this a very strange concern." Several strategies exist for dealing comfortably with nuclear waste, he pointed out. One is called "accelerator-driven transmutation," which smashes up radioactive material into more stable products with shorter half-lives, a process that actually generates more energy than it uses. Another is to employ fast-breeder reactors that use nuclear fuel over and over until all the plutonium is used up; the French are already doing this. Or reactors could use thorium instead of uranium—which is more abundant and produces less radioactive waste than uranium.[14]

The main point is simply this: nuclear power can be hazardous if mishandled, but it represents a far smaller danger to the Earth and to all our lives than business-as-usual, our continuing addiction to fossil fuels, which if we don't change quickly, will change the climate for millennia and threaten the lives of us all. We can't wait for the invention of some savior technology. We can't wait while alternative fuels and sustainable technologies go through their developments phases. The only real short- to medium-term solution is to build nuclear plants and build them fast.

3. Geo-engineering solutions

In March 2007 *The Economist* asked, in one of its quarterly technology supplements, "If man is inadvertently capable of heating the entire planet, surely it is not beyond his wisdom to cool it down?" This is the notion of geo-engineering, something that drives environmentalists up the wall—don't bother to reduce emissions, just create something even bigger and uglier, and with unforeseen consequences, to "fix" the problem, thereby almost certainly making things worse. "It's like a junkie figuring out new ways of stealing from his children," as Meinrat Andreae, an atmospheric scientist at the Max Planck Institute, put it in an interview with *Nature's* Oliver Morton.[15]

The notion of engineering our way out of the problem goes back to the 1960s, when a report to President Lyndon Johnson didn't even bother to suggest reducing emissions, instead suggesting spreading very small light-reflecting particles on the ocean surface to reflect heat and light back into space. A decade later a scientist named Mikhail Budyko suggested the Earth could be cooled by injecting tiny sunlight-reflecting particles into the stratosphere. In fact, as the journal *Science* put it, "nature soon served up a few striking examples, with the eruptions of the volcano El Chichón in 1982 and then Pinatubo in 1991. The long-lived debris from Pinatubo, water droplets laced with sulfuric acid, reflected enough sunlight to cool the Earth an average of 0.5C [33 degrees Fahrenheit] for a year or two. That's about the size of the warming of the past century."[16] The same article sent me to a more recent article in the August 2006 issue of *Climate Change*, in which Paul Crutzen, who won the Nobel Prize for helping work out the chemistry of ozone depletion, suggested deliberately creating a global-sized pollution cloud of sulfur gases and debris, which would have a similar cooling effect. "A few years ago I would have said, I'm not touching that," he said. "But I am grossly disappointed by the international political response to global warming's threat, so the notion of deliberately contaminating the air seems less crazy." In the lower atmosphere, such pollution simply hurts human lungs, but it would be just as effective a cooling agent, and longer lasting, in the stratosphere. Some models say that the resulting pollution would allow carbon dioxide levels to climb to 500 ppm without warming the Earth any further.

There have been myriad other suggestions, all of them ingenious, many of them fearfully expensive, a good many of them insane, some of them practical, almost all of them with quite unforeseeable consequences. Several conferences have been held to discuss ideas, one in January 2004 at Cambridge University, another in November 2006 by the Carnegie Institution and NASA. Then, in November 2007, a meeting was held at Harvard attended by two of the preeminent climate change scientists, James Hansen and Kerry Emanuel. Hansen was and still is against geo-engineering (his view is that proper carbon sequestration and more advanced farming could help make it unnecessary), but he went anyway. Scientists should at least understand it, he allowed, so they can head off politicians' more hare-brained implementations.

Some of the ideas emerging from these conferences:

Place a giant sunshade at the Lagrange Point, that position between the Earth and the sun where the combination of gravity and centripetal forces allows an object to remain stationary without expending energy. The proposal was made by the Lawrence Livermore Laboratory, which estimated the sunshade would weigh about 200 tons.

A similar idea, proposed by Roger Angel, an astronomer, would be to assemble a cloud of millions of small reflecting spacecraft, each less than a meter across, where they could block out 1.8 percent of the sun's rays. The total mass, Angel suggested, would be about 20 million tons, the cost a few trillion dollars or so.

Or perhaps sprinkle iron filings on a portion of the ocean, encouraging the growth of natural algae, which will absorb carbon dioxide and then, conveniently, sink when they die.

James Lovelock and Chris Rapley of Oxford suggested lowering ten-meter pipes deep into the ocean, to exchange nutrient-rich deep water for the relatively barren surface water, thereby fertilizing surface algae, causing them to bloom, thereby absorbing carbon dioxide.

John Latham, of the National Institute for Atmospheric Research in Colorado, suggests that blasting tiny droplets of seawater into the air would stimulate the formation of highly reflective low-lying marine clouds. Stephen Salter, of Edinburgh, has designed an unmanned vessel to do just that, using wind power. Just fifty vessels, he reckons, each costing a few million dollars and spraying ten kilograms of water per second, could cancel out a year's worth of global CO_2 emissions—though another fifty vessels would be needed each year.

Air streams used to power wind turbines could be scrubbed to remove CO_2. Several technological challenges remain, but what makes it possible is that air streams carry a hundred times more CO_2 than kinetic energy (or put another way, the physical dimensions of the CO_2 capture devices would be only 1 percent of the sweep of the turbines).[17]

The sequestration of carbon, either in deep mines or under the sea, is already a commonplace. Gregory Benford, a physicist at the University of California at Irvine and a novelist, proposes an ingeniously simple idea: we already "manage," through agriculture, far more carbon than is causing our greenhouse dilemma. Why not, then, take the leftover agricultural debris—corncobs and stalks from farm fields can be gathered up, "floated down the Mississippi, and dropped into the ocean, sequestering its contained carbon. . . . It's not a permanent solution, but it would buy us time to find better arguments. Sequestering crop leftovers could offset about a third of the carbon we put into our air. . . ."[18]

Canada's Alberta Research Council is even looking to turn pond scum into a profit center, re-engineering blue-green algae to absorb massive amounts of carbon dioxide.

Environmentalists oppose the whole idea of geo-engineering on principle, on the reasonable grounds that any such schemes would put at risk real solutions, such as actually doing something to reduce emissions. And pretty well everyone, including most of the proponents of such ideas, agrees that they are last-ditch fixes, to be used when everything else has failed.

Fair enough. But how do we know when everything else has failed?

Can We Do It? Will We?

[Can] angels pass from one extreme to another without going through the middle? Do angels know things more clearly in the morning? How many angels can dance on the point of a very fine needle, without jostling one another?
—Isaac d'Israeli, Benjamin Disraeli's father, quoting "Martinus Scriblerus" (probably himself) making fun of Thomas Aquinas

We are programmed by our inheritance to see other living things as mainly something to eat.
—James Lovelock, *Gaia's Revenge*

This Earth is the only place we've got. This is already a cliché. We need to cherish it, and protect it. This is a cliché too.

But *how*?

Part of the how is understanding where the boundaries are and comprehending the dangers. It is commonly said that humans are destroying the planet, but this is typical human hubris. It is ridiculous to believe that one small species, no matter what its engines of destruction, can destroy a planet that has survived four and a half billion years—and survived, at that, against a background of continuous and violent upheaval: mountain ranges rearing up and disappearing, the great chasms that are the oceans opening and closing like some gigantic flower, cometary impacts, titanic collisions, volcanic eruptions, whole continents shifting and grinding . . .

We need to get over this absurd notion that we are endangering the planet.

There have been mass extinctions in the Earth's history—and life survived, along with the planet.

Oliver Morton, features editor of *Nature*, has made the same point:

> The Earth doesn't need ice caps or permafrost or any particular sea level. Such things come and go and rise and fall as a matter of course. The planet's living systems adapt and flourish, sometimes in a way that provides negative feedback, occasionally with a positive feedback that amplifies the change. A planet that made it through the massive biogeochemical unpleasantness of the late Permian is in little danger from a doubling (or even quintupling) of the very low carbon dioxide level that preceded the Industrial Revolution, or from the loss of a lot of forests and reefs, or from the demise of half its species, or from the thinning of the ozone layer at high altitudes. . . . If fossil-fuel use goes unchecked, carbon dioxide levels may rise as high as they were in the Eocene, some 50 million years ago . . . [but] the Eocene was not such a terrible place.[1]

If the impact of an asteroid wiped out life on Earth's surface, or if a supernova radiation storm destroyed all plants and animals (both possible outcomes, as this book has shown), the Earth would survive. So would life—somewhere, perhaps as bacteria deep under the ground or in the oceans, or frozen in ice. "Life" can take strange forms, after all—microbiologist Joan Slonczewski has found organisms living one and a quarter miles below the Earth's surface, in gold mines, that feed on hydrogen atoms produced by uranium decay. As she puts it, "I have yet to see nuclear-powered creatures [in fiction, even] in science fiction" but here they are.[2] And two species of groundwater amphipods apparently made it through the last ice age in Iceland, surviving though trapped beneath the crushing weight of nearly a mile of ice. It could take a billion years or more for life to spread once more over the Earth, but spread it would.

Earth has time. But we don't. There would be life, but it wouldn't be *our* life, or even life as we more or less know it. The planet won't die, but that version of the planet that makes our existence agreeable or even possible could do so with ease. Either human-caused or natural calamities, or both in concert, could make it happen.

This is the vulnerability we need to confront and then devise policies that would maximize our chances of keeping ourselves alive and well. It remains possible. We are an inventive species as well as a destructive one. We now need to invent not just new science, but a new politics. We are doing plenty of the first. And we are beginning to do the second, with climate change the engine that's driving us. Mike Moore, who was New Zealand's prime minister before he took on the thankless job of managing the World Trade Organization, is one of the smartest political operators around. He is one of the optimists. He put it this way: "Many good people are endlessly seeking solutions for world problems. Why? Because we can no longer keep our distance from suffering. We now live the pain we see on television every night. We know the dangers of failure. Everyone is our neighbor now; their suffering degrades us all, and their success inspires us all."[3] I've heard him make the same point in speeches. Let's hope he's right.

My first draft of a postscript opened with this sentence: "The very real dangers we face as a species are the best argument for a much broader policy of human cooperation." What a resounding platitude! Even to me it sounded like the peroration of a stump speech by a losing candidate whose audience is already drifting out the door, looking for a hot dog or an ear of buttered corn or a clown or a shooting gallery to brighten their day. But the *idea* is right.

It has always been a commonplace of science fiction that what the world needs is an external enemy to bring its quarreling populations together. What few of those writers understood was that the "enemy" was not alien spaceships manned by malevolent monsters, but instead the uncaring, impersonal, value-free operations of entirely natural systems that, quite literally, surround and envelop us. Looked at this way, the political and religious quarrels that consume so much of our energy seem increasingly insane, and so does the ruinous "development" we are inflicting on our home.

The fight against global warming isn't a fight to maintain the status quo. The status quo isn't ours to keep. Change is coming. The planet adjusts, always, and so must we.

The real question is politics. If fifty million people can be killed by a tiny mutation of a microbe we can't even see, if an entire continent can be wiped out by a collision with something we can't predict, if the global climate can change when a magma chamber a few miles square suddenly decides to erupt . . . why do we spend so much of our time in fruitless quarreling? It wasn't at all reassuring that, with the American state of Georgia facing critical water shortages in the fall of 2007, one of the most vigorous political controversies was about whether the governor could legitimately pray for rain or whether that would somehow violate constitutional norms. Given the enormity of what is going on around us, surely there are higher priorities? Otherwise we might as well join the debate about whether evolution is "true" or not—of all the quarrels of our time perhaps the most deeply irrelevant, joined in its fatuity only by religious quarrels over, well, nothing—Protestant against Catholic, Old Believers versus New, Sunni versus Shia. Perhaps five angels, or fifty, can dance upon the point of a fine needle, but none of their dancing will affect the course of the tsunami that will be rolling someone's way quite soon. Or the hurricane that will be coiling its deadly way across the Caribbean this summer. Or the earthquake that will tumble down cities. Or the volcano that will spread its pall of ash and destruction across towns and villages not yet known. Or the rising sea levels that will swamp coastal communities. This is surely where our attention must be focused.

The rest of the quotation from James Lovelock that started this little polemic is this:

> We are programmed by our inheritance to see other living things as mainly something to eat, and we care more about our national tribe than anything else. We will even give our lives for it and are quite ready to kill other humans in the cruelest way for the good of our tribe. We still find alien the concept that we and the rest of life, from bacteria to whales, are parts of the much larger and diverse entity, the living Earth.[4]

Ah, but that's changing, I think. And that's our last best hope.

NOTES

1 Doomsday as a State of Mind

1. Fletcher, *Moorish Spain*, p. 79.
2. Harris, *Letter to a Christian Nation*, p. x.
3. A 2002 Time/CNN poll found that 59 percent of Americans believe that the prophecies found in the book of Revelation will come true.
4. You can force arcane meanings on many numbers, of course. Here are some that have been "found" for 666. It is the sum of the first 36 natural numbers (i.e., $1 + 2 + 3 \dots + 34 + 35 + 36 = 666$), and thus a triangular number. Since 36 is both square and triangular, 666 is the sixth number of the form $n^2(n^2 + 1) / 2$ (triangular squares) and the eighth number of the form $n(n + 1)(n^2 + n + 2) / 8$ (doubly triangular numbers). It is also the sum of the squares of the first seven prime numbers (i.e., $2^2 + 3^2 + 5^2 + 7^2 + 11^2 + 13^2 + 17^2 = 666$). And there's more: the harmonic mean of the decimal digits of 666 is an integer: $3/(1/6 + 1/6 + 1/6) = 6$, making 666 the 54th number with this property. In base 10, 666 is also a palindromic number, a redigit, and a Smith number. A prime reciprocal magic square based on 1/149 in base 10 has a magic total of 666. The Roman numeral for 666 (DCLXVI) uses only once each of the Roman numeral symbols with values under 1,000, occurring in descending order of

their respective values (D = 500, C = 100, L = 50, X = 10, V = 5, I = 1). And lastly, 666 is a member of the Indices of prime Padovan sequence, 3, 4, 5, 7, 8, 14, 19, 30, 37, 84, 128, 469, 666, 1262, 1573, 2003, 2210.

5. Hitchens, *God Is Not Great: How Religion Poisons Everything*, p. 84.

6. "News in Brief," *Nature* 421, 882, 27 February 2003.

7. Deena Beasley, "British Scientist Puts Odds for Apocalypse at 50–50," 10 June 2003 © Reuters. www.commondreams.org/headlines03/0609-07.htm. Accessed 25 October 2007.

8. Fisher, *Fire and Ice: The Greenhouse Effect, Ozone Depletion and Nuclear Winter*, p. 34.

9. Malcolm Brown, *The New York Times*, 10 August 1999.

10. Brown, op. cit.

11. Kevin Krajick, "Tracking Myth to Geological Reality," *Science* 310, 4 November 2005, pp. 762–64.

12. If you want a good explanation for a layperson, try Nick Bostrom, director of the Future of Humanity Institute in the Faculty of Philosophy at Oxford University, the author of *Anthropic Bias: Observation Selection Effects in Science and Philosophy* (New York: Routledge, 2002). His web page is www.nickbostrom.com.

13. Jim Holt, "Doom Soon: A Philosophical Invitation to the Apocalypse," *Lingua Franca* 7 (8), October 1997.

14. Leslie has taken up the Doomsday Argument in his monograph *The End of the World* (Routledge, 1996).

15. See *Stanford Encyclopedia of Philosophy*.

16. Of course, there are ways around this logical impasse. If you take as your universe of people to be affected only those alive at the time of the potential catastrophe (on the grounds that a calamity can't kill your grandfather because he's already dead), then the numbers alive now and the numbers alive in the future are not very different (the Earth has a limited carrying capacity). Therefore the odds of extinction now and then are pretty similar.

17. Brockman (ed.), *What Is Your Dangerous Idea?*, p. 4.

18. Max Tegmark and Nick Bostrom, "Astrophysics: Is a Doomsday Catastrophe Likely?" *Nature* 438, 8 December 2005.

2 Catastrophe in Human Life: The Probability Theorem

1. Keys, *Catastrophe: An Investigation into the Origins of the Modern World*, p. 3.

2. Keys, *Catastrophe*, p. 239.

3. Keys, *Catastrophe*, p. 246.

4. Stanley H. Ambrose, "Late Pleistocence Human Population Bottlenecks, Volcanic Winter, and Differentiation of Modern Humans," *Journal of Human Evolution* 34, 1998, pp. 623–51. The opposing Gathorne-Hardy and Harcourt-Smith paper, entitled "The Super-eruption of Toba: Did It Cause a Human Bottleneck?" was published in the same journal, issue 45, pp. 227–30.

5. Some of this is from a website operated by consultant Peter Webb (http://www.probabilitytheory.info/) and some from Sherman DeForest on lockergnome.com.
6. Webb, op. cit.
7. Quirin Schiermeier, "Natural Disasters: The Chaos to Come," *Nature* 438, 15 December 2005, p. 903.

3 Our Perilous Neighborhood: Understanding Cosmology

1. Mark Buchanan, "Many Worlds: See Me Here, See Me There," *Nature* 448, 5 July 2007, pp. 15–17.
2. Leon Lederman, *The God Particle et al.*, in *Nature* 448, 19 July 2007.
3. Of course, it's only called the Milky Way in the Eurocentric imagination, so named after a Greek goddess, Hera, who sprayed milk across the sky in a fit of godly pique (the myth has since been neatly incorporated into Christian history, since Charlemagne followed the "voie lactée" to Compostella in Spain, thereby coming across the tomb of the apostle St. James, giving him a really good excuse to attempt the ejection of the heretic "Moors" from Spain). When I was growing up we called it the Heavenly Street (in Afrikaans). The Chinese call it the Silver River. Better yet is the more poetic and less prosaic name given the galaxy by the San of the Kalahari Desert, who call it the Backbone of Night.
4. Lunar and Planetary Institute (http://www.lpi.usra.edu/), a NASA-funded institute in Houston, Texas.

4 This Plastic Earth: Tectonics and Wandering Continents

1. Olduvai passages reprised from *Into Africa,* de Villiers and Hirtle.
2. *Le Figaro*, 14 December 2005.
3. Ting-Li Lin and Charles A. Langston, "Infrasound from Thunder: A Natural Seismic Source," *Geophysical Research Letters*, 334, L14304 (published 18 July 2007).
4. Richard Gross and Benjamin Chao, NASA scientists, in *Eos*, newspaper of the American Geophysical Union, January 2005.
5. Richard A. Kerr, "Earth's Inner Core Is Running a Tad Faster Than the Rest of the Planet," *Science* 309, p. 1313.
6. Fortey, *The Earth: An Intimate History*, p. 414.
7. Alexandra Witze, "Geology: The Start of the World as We Know It," *Nature* 442, 13 July 2006.
8. Fortey, *The Earth*, p. 190.
9. Laura Wallace et al., "Research Highlights," *Nature* 438, 24 November 2005, p. 399.
10. Fortey, *The Earth*, p. 434.

5 Our Ever-Changing Climate: Ice Ages Now and Then

1. Lovelock, *The Revenge of Gaia*, pp. 42–43.
2. Stuart Clark, "The Dark Side of the Sun," *Nature* 441, 25 May 2006.

3. Novosti News Agency interview, 2005.

4. Robert Ehrlich, "Solar Resonant Diffusion Waves as a Driver of Terrestrial Climate Change," *Astrophysical Papers*, arXiv: astro-ph/0701117v1, 4 January 2007.

5. Nir J. Shaviv, "Cosmic Ray Diffusion from the Galactic Spiral Arms, Iron Meteorites and a Possible Climatic Connection," *Physical Review Letters* 89, 29 July 2002.

6. James Zachos et al., "Trends, Rhythms, and Aberrations in Global Climate 65 Ma to Present," *Science* 292, 27 April 2001.

7. See K.J. Willis et al., "The Role of Sub-Milankovitch Climatic Forcing in the Initiation of the Northern Hemisphere Glaciation," *Science* 285, 1999, p. 564.

8. Lawrence Solomon, "Forget Warming—Beware the New Ice Age," *National Post*, 25 June 2007.

6 Fragile Life: The Conundrum of Mass Extinctions

1. Nick Lane, "Reading the Book of Death," *Nature* 448, 12 July 2007.

2. David Cyranoski, "Indonesia Eruption: Muddy Waters," *Nature* 445, 22 February 2007.

3. Alvarez, *T. Rex and the Crater of Doom*, p. 57.

4. A. Hildebrand et al., "Chicxulub Crater: A Possible Cretaceous/Tertiary Boundary Impact Crater on the Yucatan Peninsula, Mexico," *Geology* 19, 1991, pp. 867–71.

5. Late in 2007, a study by William Bottke of the Southwest Research Institute in Boulder, Colorado, suggested that a "family" of asteroids called Baptistima had collided some 160 million years ago and spun off great chunks in erratic orbits. More than 100 million years of wandering later, one of Baptistima's daughters crashed into the Earth, profoundly rattling our planet and wiping out the dinosaurs. Reported in *Nature,* 6 September 2007 ("An Asteroid Breakup 160 Myr Ago as the Probable Source of the K/T Impactor").

6. *Los Angeles Times*, 5 December 1990.

7. Olaf Bininda-Emonds et al., "The Delayed Rise of Present-Day Mammals," *Nature* 446, 29 March 2007.

8. Jan A. van Dam et al., "Long-Period Astronomical Forcing of Mammal Turnover," *Nature* 443, 12 October 2006.

9. Their formal paper was published in *Proceedings* of the National Academy of Sciences, 27 September 2007.

7 The Perils Without: Comets and Asteroids

1. *The Economist*, 23 December 2006.

2. Zebrowski, *Perils of a Restless Planet*, p. 217.

3. Other sponsors included the Aerospace Corporation; Space Studies Institute; SpaceWorks Engineering Inc.; The Planetary Society; Applied Physics Laboratory; Ball Aerospace and Technologies Corporation; Orbital Sciences Corporation; and General Dynamics.

4. The Aerospace Corporation, "General Information: 2007 Planetary Defense Conference," www.aero.org/conferences/planetarydefense.

5. Zebrowski, *Perils of a Restless Planet*, p. 217.

6. D. Nesvorny, W.F. Bottke, L. Dones, and H. Levison, "A Recent Asteroid Breakup in the Main Belt," *Nature* 417, 2002, pp. 700-22.

7. J.D. Giorgini et al., "Asteroid 1950 DA's Encounter with Earth in 2880: Physical Limits of Collision Probability Prediction," *Science* 5 April 2002, pp. 132–36.

8. Guy Gugliotta, "Science's Doomsday Team vs. the Asteroids," *The Washington Post*, 9 April 2005, p. A01.

9. Govert Schilling, "And Now, the Asteroid Forecast . . ." *Science* 285, 30 July 1999.

10. Torino Scale copyright Richard P. Binzel, Massachusetts Institute of Technology.

11. Gugliotta, "Science's Doomsday Team vs. the Asteroids."

12. In the United States and Canada, you are much more likely to die of a heart attack or cancer (lifetime odds: 1 in 5 and 1 in 7) than any natural calamity. Statistics compiled by Robert Roy Britt (livescience.com) from a variety of sources, including the National Center for Health Statistics, the CDC, American Cancer Society, National Safety Council, International Federation of Red Cross and Red Crescent Societies, World Health Organization, U.S. Geological Survey, and others give the lifetime odds of U.S. residents dying this way:

Motor Vehicle Accident	1 in 100
Suicide	1 in 121
Assault by Firearm	1 in 325
Drowning	1 in 8,942
Air Travel Accident	1 in 20,000
Flood	1 in 30,000
Tornado	1 in 60,000
Lightning Strike	1 in 83,930
Snake, Bee, or Other Venomous Bite or Sting	1 in 100,000
Earthquake	1 in 131,890
Asteroid Impact	1 in 500,000
Tsunami	1 in 500,000

13. Patricia A. Lockridge et al., "Tsunamis and Tsunami-like Waves of the Eastern United States," *Science of Tsunami Hazards* 20 (3), 2002, p. 120.

8 Earthquakes

1. Fortey, *The Earth*, p. 317.

2. Adapted from Zebrowski, *Perils of a Restless Planet*, p. 179.

3. Herb Dragert, "Mediating Plate Convergence," *Science* 315, 26 January 2007, pp. 471–72.

4. Brennan Clarke, "Earthquake Scientists Detect Plate Slippage," *Victoria News*, 2 February 2007.

5. Cecep Subarya et al., "Plate-boundary Deformation Associated with the Great Sumatra-Andaman Earthquake," *Nature* 440, 2 March 2006.

6. Nicole Feldl and Roger Bilham, "Great Himalayan Earthquakes and the Tibetan Plateau," *Nature* 444, 9 November 2006.

7. Suzanne Leroy et al., "Are an Early Byzantine Seismic Event (Recorded in Lake Manyas Sediment, N-W Turkey) and the End of the Beysehir Occupation Phase Linked?" Extended abstract for NATO workshop, Belgium, May 2003.

8. Ross Stein et al., "A New Probabilistic Seismic Hazard Assessment for Greater Tokyo," *Philosophical Transactions of the Royal Society Series A* 364(1845), August 2006.

9. Winchester, *A Crack in the Edge of the World*, p. 23.

10. Winchester, *A Crack in the Edge of the World*, p. 103.

9 Volcanoes

1. Some of these numbers are from vulcanologist Dr. Stanley Williams, professor of geology at Arizona State University, who is perhaps best known for his work on volcanic predictors.

2. Richard V. Fisher, University of California at Santa Barbara, Geology Department, http://volcanology.geol.ucsb.edu/. Fisher's book *Volcanoes: Crucibles of Change* has been influential in shaping the study of volcanoes.

3. P.J. Gleckler et al., "Volcanoes and Climate: Krakatoa's Signature Persists in the Ocean," *Nature* 439, 9 February 2006.

4. This story has been recounted in many places. The most evocative is that in Scarth, *Vulcan's Fury*, p. 191.

5. Scarth, *Vulcan's Fury*, p. 157.

6. Fortey, *The Earth*, p. 7.

7. Fortey, *The Earth*, p. 29.

8. Translated by John Dryden.

9. http://www.santorini.net/caldera.html.

10. A good description of the catastrophe can be found in Zebrowski, *Perils of a Restless Planet*, pp. 6–13, from which these data are taken.

11. "Random Samples," *Science* 311, 10 March 2006.

12. Zebrowski, *Perils of a Restless Planet*, p. 213.

13. Charles W. Wicks et al., "Uplift, Thermal Unrest and Magma Intrusions at Yellowstone Caldera," *Nature* 440, 2 March 2006.

14. Wicks et al., "Uplift, Thermal Unrest and Magma Intrusions at Yellowstone Caldera."

10 *Poisonous emissions and noxious gases*

1. The webcam is at http://perso.orange.fr/mhalb/nyos/webcam.htm. On the same site you can find links to Halbwachs's indignant diatribe against American scientists he claims are trying to take credit for the project, with which (in his view) they had nothing to do. "The real truth," he writes, "is that this huge task was accomplished by more than 12 French scientists, engineers and technicians from various disciplines who worked enthusiastically for nearly 20 years on the project. For these guys it is amazing that anyone would want to steal their thunder! Or their water spout!"

2. Sean Nee, reviewing a book about Archaea in *Nature* 448, 26 July 2007, p. 413.

3. Richard A. Kerr, "A Smoking Gun for an Ancient Methane Discharge," *Science* 286, 19 November 1999.

4. Allen M. Gontz et al., "Current and Proposed Pockmark Research in Coastal Maine: Life Cycles and Fluid Origins," *Geological Society of America* paper No. 187-0. Also Silver Donald Cameron, Halifax *Chronicle Herald*, 2 September 2007.

5. William Dillon and Keith Kvenvolden, "Gas (Methane Hydrates)—A New Frontier," U.S. Geological Survey, Marine and Coastal Geology Program, http://marine.usgs.gov/fact-sheets/gas-hydrates/title.html, September 1992.

11 *Tsunamis*

1. On 10 June 1996, in the Andreanov Islands 80 kilometers southwest of Adak, Alaska, after a magnitude 7.7 earthquake occurred; on 7 July 1998 in Papua New Guinea, after a magnitude 7.1 earthquake produced 33-foot wave trains that killed more than 3,000 people, scouring away whole villages; on 26 November 1999, in the Vanuatu Islands in the western Pacific, a magnitude 7.5 earthquake generated a tsunami that attacked the coast of Vanuatu, with runups of more than 16 feet (Baie Martelli, a village near the southern tip of Pentecost, was completely destroyed by the waves); on 13 January 2001 in El Salvador, a 7.9 earthquake generated a tsunami that did little damage itself (the earthquake killed several hundred people); on 23 June 2001, a shallow (9 kilometers) 8.4 earthquake just off Peru produced a 8.43-foot tsunami that was recorded throughout the Pacific basin, still at half a meter as far away as New Zealand and Japan; on 25 September 2003, a magnitude 8.3 earthquake struck Hokkaido, in Japan; it was relatively deep (greater than 12 miles) and its tsunami, striking the rugged and sparsely inhabited coastline, killed no one. Still, 589 people were injured and there was significant damage to coastal communities. On 26 December 2004, the so-called Christmas Day earthquake and tsunami was the most destructive tsunami event ever; on 28 March 2005, an 8.2 earthquake struck the Indonesian island of Nias, killing some 2,000 people; the resulting tsunami was only moderate, and did little damage; on 17 July 2006, a magnitude 7.7 quake struck off the coast of Java, causing a 10-foot tsunami that killed 668 people, destroying several villages;

on 15 November 2006, an 8.3 earthquake struck the Kuril Islands, but generated only a one-and-a-half-foot tsunami.

2. The others, in descending order of awfulness, are these (casualties, date, magnitude of causative earthquake, location): 100,000-plus dead, 1410 BC, magnitude unknown, Crete-Santorini; 60,000 dead, 1755, 8.5, Portugal and Morocco; 40,000 dead, 1782, 7.0 probable, South China Sea; 36,500 dead, 1883, volcanic origin, Krakatoa; 30,000 dead, 1707, 8.4 estimated, Tokaido-Nankaido, Japan; 26,360 dead, 1896, 7.6, Sanriku, Japan; 25,674 dead, 1868, 8.5, Northern Chile; 15,030 dead, 1792, 6.4, Kyushu Island, Japan; 13,486 dead, 1771, 7.4, Ryukyu Trench, Japan.

3. Zebrowski, *Perils of a Restless Planet*, p. 151.

4. For the technically minded, the wave speed is the square root of the product of the gravity constant (g) and water depth. The calculations in the chart are from Georgia State University's physics department, at http://hyperphysics.phyastr.gsu.edu/hbase/waves/tsunami.html.

5. Brian F. Atwater et al., *Surviving a Tsunami—Lessons from Chile, Hawaii, and Japan*, http://pubs.usgs.gov/circ/c1187/.

6. Zebrowski, *Perils of a Restless Planet*, pp. 132–35.

7. Alan Ruffman, "Comment On: Tsunamis and Tsunami-Like Waves of the Eastern United States," *Science of Tsunami Hazards* 23(3), p. 52.

8. Jacopo Pasotti, "Tracking a Killer Tsunami," *Science* 314, 8 December 2006.

9. J.L. Moss et al., "Ground Deformation Monitoring of a Potential Landslide at La Palma, Canary Islands," *Journal of Volcanology and Geothermal Research* 94, pp. 251–65.

10. N. Driscoll et al., "Potential for Large-Scale Submarine Slope Failure and Tsunami Generation Along the U.S. Mid-Atlantic Coast," *Geology* 28(5), May 2000, pp. 407–10.

11. Much of this discussion of Atlantic tsunamis is from Patricia A. Lockridge, "Tsunamis and Tsunami-like Waves of the Eastern United States," referred to earlier in Chapter 7. Lockridge also led me to McGuire and others.

12. Dunn and Miller, *Atlantic Hurricanes*.

12 Floods

1. Ryan and Pitman's results were published in their 1998 book, *Noah's Flood: The New Scientific Discoveries About the Event That Changed History* (reissued by Simon and Schuster in paperback in 2000). Their conclusions have been widely reported, including in a PBS documentary and dozens of news reports. The Abrajano material is from Rensselaer Polytechnic Institute, 17 June 2002, "Noah's Flood Hypothesis May Not Hold Water," *Science Daily*, 20 November 2007 (www.sciencedaily.com). The Ballard confirmatory expedition was sponsored by National Geographic society.

2. Ice Age Floods Institute, http://www.iafi.org/index.htm.

3. J.A. Rayburn et al., "Evidence from the Lake Champlain Valley for a Later Onset of the Champlain Sea and Implications for Late Glacial Meltwater Routing to the North Atlantic," *Palaeogeography, Palaeoclimatology, Palaeoecology* 246, pp. 62–74.

4. Sanjeev Gupta et al., "Catastrophic Flooding Origin of Shelf Valley Systems in the English Channel," *Nature* 448, 19 July 2007, pp. 342–45.

5. Much of this information is sourced from "The Great Flood of Summer 1993: Mississippi River Discharge Studied," *Earth in Space* 7(3), November 1994, pp. 11–14.

6. Details and quotes from *The Washington Post* piece ("La. Plan to Reclaim Land Would Divert the Mississippi") by Peter Whoriskey, 1 May 2007.

7. Carlo del Ninno et al., "The 1998 Floods in Bangladesh: Disaster Impacts, Household Coping Strategies, and Responses," Research Reports 122, International Food Policy Research Institute (IFPRI), 2001.

8. Frank Wentz et al., "How Much More Rain Will Global Warming Bring?" *Science* 317, 13 July 2007, pp. 233–35.

9. The report can be found at http://www.rms.com/publications/1607_bristol _flood.pdf.

13 Vile Winds: Tropical Cyclones and Tornadoes

1. The Fujita tornado scale:

Fujita 0, gale tornado: Winds 40 to 71 miles an hour. Such tornadoes cause some damage to chimneys, break branches off trees, and push over shallow-rooted trees.

Fujita 1, moderate tornado: Winds ranging from 72 to 112 miles an hour. The lower limit of a moderate tornado is the sustained wind speed that defines a Category 1 hurricane. Such tornadoes can peel off roofs, overturn mobile homes, and push cars off roads. Some poorly made buildings will be destroyed.

Fujita 2, significant tornado: Winds ranging from 113 to 180 miles an hour. Such winds will do considerable damage, tearing roofs off many houses, demolishing mobile homes, snapping large trees. "Light-object missiles" will be generated—debris picked up in the winds that become battering rams.

Fujita 3, severe tornado: Winds from 181 to 206 miles an hour. Roofs and walls torn off well-made buildings, trees uprooted, and even trains overturned.

Fujita 4, devastating tornado: With winds ranging from 207 to 260 miles an hour. In these conditions even well-made houses are leveled. Structures with weak foundations will be blown some distance. Cars are thrown about and "large missiles" generated.

Fujita 5, incredible tornado: Winds of 261 to 318 miles an hour. Strong frame houses lifted off their moorings, car-sized missiles flying about, trees debarked, steel-reinforced concrete badly damaged.

Fujita 6, inconceivable tornado: Sustained winds of 319 to 379 miles an hour, but no one will ever know, because all measuring devices would be destroyed, along with pretty well everything else. (The Fujita scale recognizes that "the small area of damage they might produce would probably not be recognizable along with the mess produced by F4 and F5 wind that would surround the F6 winds. Missiles, such as cars and refrigerators would do serious secondary damage that could not be directly identified as F6 damage. If this level is ever achieved, evidence for it might only be found in some manner of ground swirl pattern, for it may never be identifiable through engineering studies."

2. Source: Munich Reinsurance Company. Press release December 2004, and bulletin, *Hurricanes—More Intense, More Frequent, More Expensive: Insurance in a Time of Changing Risks*, 2006.

3. The Saffir-Simpson Hurricane Scale:

Category 1: Sustained winds 74 to 94 miles an hour (64 to 82 knots). Storm surge generally 3.9 to 4.9 feet above normal. No real damage to building structures. Damage primarily to unanchored mobile homes, shrubbery, and trees. Some damage to poorly constructed signs. Also, some coastal road flooding and minor pier damage.

Category 2: Sustained winds of 95 to 110 miles an hour (83 to 95 knots). Storm surge generally 6.5 to 8.2 feet above normal. Some roofing material, door, and window damage. Considerable damage to shrubbery and trees with some trees blown down. Considerable damage to mobile homes, poorly constructed signs, and piers. Coastal and low-lying escape routes flood 2 to 4 hours before arrival of the hurricane center. Small craft in unprotected anchorages break moorings.

Category 3: Sustained winds 111 to 130 miles an hour (96 to 113 knots). Storm surge generally 8.9 to 11.8 feet above normal. Some structural damage to small residences and utility buildings with a minor amount of curtainwall failures. Damage to shrubbery and trees with foliage blown off trees and large trees blown down. Mobile homes and poorly constructed signs are destroyed. Low-lying escape routes cut by rising water 3 to 5 hours before arrival of the center of the hurricane. Flooding near the coast destroys smaller structures with larger structures damaged by battering from floating debris. Terrain lower than 5 feet above sea level may be flooded inland 7.5 miles or more. Evacuation of low-lying residences within several blocks of the shoreline may be required.

Category 4: Sustained winds of 131 to 154 miles an hour (114 to 135 knots). Storm surge generally 12.8 to 17.7 feet above normal. More extensive curtainwall failures with some complete roof structure failures on small residences. Shrubs, trees, and all signs blown down. Complete destruction of mobile homes. Extensive damage to doors and windows. Low-lying escape routes may be cut by rising water 3 to 5 hours before arrival of the center of the hurricane. Major damage to lower floors of structures near the shore. Terrain lower than 10 feet above sea level may be flooded, requiring massive evacuation of residential areas as far inland as 6 miles.

Category 5: Sustained winds greater than 155 miles an hour (135 knots). Storm surge generally greater than 17.7 feet above normal. Complete roof failure on many residences and industrial buildings. Some complete building failures with small utility buildings blown over or away. All shrubs, trees, and signs blown down. Complete destruction of mobile homes. Severe and extensive window and door damage. Low-lying escape routes are cut by rising water 3 to 5 hours before arrival of the center of the hurricane. Major damage to lower floors of all structures located less than 15 feet above sea level and within 1,640 feet of the shoreline. Massive evacuation of residential areas on low ground within 5 and 10 miles of the shoreline may be required.

4. Some of this discussion about hurricanes was first published in de Villiers, *Windswept* (Walker and Co. in the United States; McClelland and Stewart in Canada).

14 Plague and Pandemic

1. Some of this from de Villiers and Hirtle, *Into Africa*.
2. AP, "Animal Diseases Jumping to Humans," Halifax *Chronicle Herald*, 20 February 2006.
3. N.D. Wolfe et al., "Naturally Acquired Simian Retrovirus Infections in Central African Hunters," *The Lancet* 363, 20 March 2004, p. 932.
4. Nathan Wolfe et al., "Origins of Major Human Infectious Diseases," *Nature* 447, 17 May 2007.
5. Howard Markel, "Old Plagues Don't Die, They Evolve," *The Washington Post*, 10 June 2007.
6. Ian Pindar, "Round the World on a Rat," *The Guardian*, 25 May 2007, reviewing William Rosen's book *Justinian's Flea*, Rosen in turn quoting Procopius.
7. Thomas Maugh, "An Empire's Epidemic: Scientists Use DNA in Search for Answers to 6th Century Plague," *Los Angeles Times*, 6 May 2002.
8. This account of Justinian's Plague, and the Black Death that follows, are drawn from a variety of sources, including *Plague and the End of Antiquity: The Pandemic of 750*, various authors, by Cambridge University Press; *The Black Death* by Ole Benedictow; and William Rosen's *Justinian's Flea: Plague, Empire and the Birth of Europe*. Also useful was a piece by Carroll Payne ("The Effects of the Black Death on Peasant Uprisings and Persecutions") in *World Conflict Quarterly*, February 2002, and the CDC website (www.cdc.gov).
9. Recounted on the US government's pandemic website (www.pandemicflu.gov).
10. Cantor, *In the Wake of the Plague*, p. 181 and p. 12.
11. Maugh, op. cit.
12. CDC website.
13. Hemagglutinin and neuraminidase.

14. From www.pandemicflu.gov.
15. These anecdotes, and the children's rhyme, from Molly Billings, for Stanford University (http://virus.stanford.edu/uda/), 2005.
16. N. Ferguson, "Poverty, Death and a Future Influenza Epidemic," *The Lancet* 358(9553), 21 December 2006.

15 Making Things Worse: Acts of God and Acts of Man

1. Zebrowski, *Perils of a Restless Planet*, p. 109.
2. Martin, *The Meaning of the 21st Century*, p. 53.
3. Juliette Jowit, "No One Is Willing to Address the Accelerating Growth in the World's Population," *The Guardian Weekly*, 23 March 2007.
4. Quoted by Kolbert in *Field Notes from a Catastrophe*, p. 143.
5. Kurzweil, "The Near-Term Inevitability of Radical Life Extension and Expansion," in Brockman (ed.), *What Is Your Dangerous Idea*, p. 215.
6. Daly and Townsend (eds.), *Valuing the Earth*, p. 267.
7. Environmental News Service, "First Short-Term Global Warming Forecast: Record Heat," August 2007.
8. Fisher, *Fire and Ice*, p. 64.
9. F.J. Wentz et al., "How Much More Rain Will Global Warming Bring?" *Science* 317, 13 July 2007.
10. Hansen's study was available through sciencexpress, 28 April 2005 (www.sciencexpress.org).
11. News item, *Science* 315, 5 January 2007.
12. Quirin Schiermeier, "No Solar Hiding Place for Greenhouse Sceptics," *Nature* 448, 5 July 2007, pp. 8–9.
13. Leslie, *The End of the World*, p. 33.
14. Details from Thomas V. Inglesby et al., "The Hemorrhagic Fever Viruses as Biological Weapons: Medical and Public Health Management," a report in the *Journal of the American Medical Association* (Vol. 287, 2002, pp. 2391–405) on the prospect of plague as a biological weapon.
15. News item, *Nature* 317, 28 September 2007.
16. Ando Arike, "Owning the Weather: The Ugly Politics of the Pathetic Fallacy," *Harper's*, January 2006, p. 67.
17. C. Venkatamaran et al., Indian Institute of Technology, news item, *Science*, 4 March 2005.
18. V. Ramanathan et al., "Warming Trends in Asia Amplified by Brown Cloud Solar Absorption," *Nature* 448, 2 August 2007.
19. Environmental News Service, 14 August 2007.

16 Making Things Better (i): Mitigating Natural Calamities

1. Quirin Schiermeier, "Natural Disasters: The Chaos to Come," *Nature* 438, 15 December 2005.
2. Zebrowski, *Perils of a Restless Planet*, p. 30.

3. Both these ideas were contained in an interesting piece ("Disasters: Searching for Lessons from a Bad Year") by John Bohannon in *Science* 310, 23 December 2005, p. 1883.

4. For an interesting discussion of the precautionary principle, see Sunstein, *Laws of Fear*, p. 4, where this idea is cogently argued.

5. Zebrowski, *Perils of a Restless Planet*, p. 53.

6. Charles Miller, "Hurricane Warnings," *Fine Homebuilding*, December 1992/ January 1993, p. 82.

7. A diverting discussion on this effort can be found in a piece ("Riders on the Storm") by Graeme Wood in *The Atlantic*, October 2007; a fuller account of weather mitigation attempts can be found in my own *Windswept*.

8. From Peter Edidin, "Afloat in the Flood Zone," *The New York Times*, 27 October, 2005.

9. Halifax *Chronicle Herald*, 20 February 2006 (AP story).

10. The H5N1 virus is immune to amantadine and rimantadine, two antiretrovirals currently used to treat flu, but as of 2007 two others—oseltamavir and zana-mavir—were working.

11. Ran Balicer, "Modeling Infectious Diseases Dissemination Through Online Role-Playing Games. Virtual Epidemiology," *Epidemiology* March 2007, pp. 260–61.

12. M. Galimand et al., "Multidrug Resistance in *Yersinia pestis* Mediated by a Transferable Plasmid," *New England Journal of Medicine* 337, 4 September 1997, pp. 677–81.

17 *Making Things Better (ii): Undoing Human-Made Calamities*

1. "European Climate Change Programme II: Impacts and Adaptation," http://ec.europa.eu/environment/climat/eccp_impacts.htm, as reported by Environmental News Service (ENS).

2. Robert Costanza, "Embodied Energy and Economic Valuation," *Science* 210, 12 December 1980.

3. Mill, *Principles of Political Economy*.

4. Daly and Townsend (eds.), *Valuing the Earth*, p. 374.

5. Richard Seager et al., "Model Projections of an Imminent Transition to a More Arid Climate in Southwestern North America," *Science* 316, 25 May 2007, pp. 1181–84.

6. Declan Butler, "Energy Efficiency: Super Savers: Meters to Manage the Future," *Nature*, 8 February 2007.

7. Declan Butler, "Plug in, Turn on . . . Sell Out," *Nature*, 8 February 2007.

8. Lydia Polgreen, "In Niger, Trees and Crops Turn Back the Desert," *The New York Times*, 12 February 2007.

9. Alana Herro, "New Solar Cell Most Efficient Ever," worldwatch.org 20 December 2006, and Worldwatch journal, March–April 2007.

10. Newton, *Encyclopedia of Air*, p. 228.

11. His report, "Renewable and Nuclear Heresies," appeared in the *International Journal of Nuclear Governance, Economy and Ecology* 1(3), 2007.
12. Lovelock, *The Revenge of Gaia*, p. 100.
13. Lovelock, *The Revenge of Gaia*, p. 91.
14. Al-Khalili is also professor of public engagement at the University of Surrey. These quotes are from a piece ("Nuclear Waste Is Hardly a Worry When the Climate Change Threat Is So Urgent") he published in *The Guardian Weekly*, 3 August 2007.
15. Oliver Morton, "Climate Change: Is This What It Takes to Save the World?" *Nature* 447, 10 May 2007.
16. Richard A. Kerr, "Pollute the Planet for Climate's Sake," *Science* 314, 20 October 2006.
17. Wallace S. Broecker, "CO_2 Arithmetic," *Science* 315, 9 March 2007.
18. Benford, "Think Outside the Kyoto Box," in Brockman (ed.). *What Is Your Dangerous Idea?*, p. 45.

Postscript: Can We Do It? Will We?

1. Morton, "Our Planet Is Not in Peril," in Brockman (ed.), *What Is Your Dangerous Idea?*, p. 51.
2. Joan Slonczewski , "The Biologists Strike Back," *Nature* 448, 5 July 2007.
3. Mike Moore, *A World Without Walls*, p. 42.
4. Lovelock, *The Revenge of Gaia*, p. 4.

BIBLIOGRAPHY

The following books are all worthy of a reader's further attention. But for me, three of the authors stand out. The first is Richard Fortey, a geologist with an engaging manner, a poet's eye, and a lovely way with metaphor. The second is Ernest Zebrowski, professor of physics at Pennsylvania State University's College of Technology and also founder of the doctoral program in science and math education at Southern University, a historically black university in Baton Rouge, Louisiana. Zebrowski is a scientist, but he writes with a clarity of prose that George Orwell would have envied. His *Perils of a Restless Planet* covers much of the same ground as the present volume, written from a scientific perspective. His more recent book, on Hurricane Camille and its aftermath, is to me the best book yet done on the successes and failures in dealing with violent tropical cyclones. The third is the economist Herman Daly, whose thoughts on where we go from here and on how we should organize our economies seem to me as worthy of attention as those of anyone writing about the natural world itself.

That two of my own earlier books make it to this list is a somewhat lesser matter. They are cited here mostly because certain passages in this volume are amplifications of thoughts suggested in those earlier works.

The many technical articles and papers that were essential for researching this book are cited in the chapter notes, and it seems redundant to repeat them here.

Alvarez, Walter. *T. Rex and the Crater of Doom*. New York: Vintage/Random House, 1998.

Anderson, Mary. *Do No Harm: How Aid Can Support Peace—Or War*. Boulder, Col.: Lynne Rienner Publishers, 1999.

Bakker, Robert T. *The Dinosaur Heresies: New Theories Unlocking the Mystery of the Dinosaurs and Their Extinction*. New York: William Morrow, 1986.

Benedictow, Ole J. *The Black Death*. Woodbridge, UK: Boydell & Brewer Inc., 2006.

Benton, Michael J. *When Life Nearly Died: The Greatest Mass Extinction of All Time*. London: Thames and Hudson, 2003.

Brockman, John, ed. *What Is Your Dangerous Idea?* New York: Harper Perennial, 2007.

Bryson, Bill. *A Short History of Nearly Everything*. Toronto: Anchor Canada, 2004.

Cantor, Norman F. *In the Wake of the Plague*. New York: Harper Perennial, 2002.

Daly, Herman, and Kenneth Townsend, eds. *Valuing the Earth: Economics, Ecology, Ethics*. Cambridge, Mass.: MIT Press, 1993.

de Villiers, Marq, and Sheila Hirtle. *Into Africa: A Journey Through the Ancient Empires*. Toronto: Key Porter Books, 1997.

de Villiers, Marq. *Windswept: The Story of Wind and Weather*. New York: Walker Books & Toronto: McClelland and Stewart, 2006.

Dunn G.E., and B.I. Miller. *Atlantic Hurricanes*. Baton Rouge, La.: Louisiana State University Press, 1964.

Erwin, Douglas. *Extinction: How Life on Earth Nearly Ended 250 Million Years Ago*. Princeton: Princeton University Press, 2006.

Fisher, David E. *Fire and Ice: The Greenhouse Effect, Ozone Depletion and Nuclear Winter*. New York: Harper and Row, 1990.

Fletcher, Richard. *Moorish Spain*. Berkeley, Calif.: University of California Press, 2006.

Fortey, Richard. *The Earth: An Intimate History*. London: HarperCollins, 2004.

Gleick, James. *Isaac Newton*. New York: Vintage, 2003.

Harris, Sam. *Letter to a Christian Nation*. New York: Alfred A Knopf, 2006.

Hitchens, Christopher. *God Is Not Great: How Religion Poisons Everything*. Toronto: McClelland and Stewart, 2007.

Keys, David. *Catastrophe: An Investigation into the Origins of the Modern World.* New York: Ballantine, 1999

Kolbert, Elizabeth. *Field Notes from a Catastrophe.* London: Bloomsbury, 2006.

Leslie, John. *The End of the World.* London: Routledge, 1996.

Little, Lester K., ed. *Plague and the End of Antiquity: The Pandemic of 541–750.* Cambridge: Cambridge University Press, 2007.

Lovelock, James. *The Revenge of Gaia.* London: Allen Lane, 2006.

Martin, James. *The Meaning of the 21st Century.* New York: Riverhead Books, 2006.

J.S. Mill. *Principles of Political Economy,* vol. 2. London: John W. Parker and Son, 1857.

Moore, Mike. *A World Without Walls.* Cambridge: Cambridge University Press, 2003.

Newton, David E. *Encyclopedia of Air.* Westport, Conn.: Greenwood Press, 2003.

Preston, Douglas. *Tyrannosaur Canyon.* New York: Forge, 2005.

Rosen, William. *Justinian's Flea: Plague, Empire and the Birth of Europe.* New York: Simon & Schuster, 2004.

Rutherford, R.J., and H. Breeze. *The Gully Ecosystem.* Canadian Manuscript Report of Fisheries and Aquatic Sciences 2615. Fisheries and Oceans Canada.

Ryan, W.B., and W.C. Pitman. *Noah's Flood: The New Scientific Discoveries About the Event That Changed History.* New York: Touchstone, 2000.

Scarth, Alwyn. *Vulcan's Fury: Man Against the Volcano.* New Haven: Yale University Press, 1999.

Schatzing, Frank. *The Swarm.* Translated by Sally-Ann Spencer. New York: HarperCollins, 2006.

Scott, Susan, and Christopher Duncan. *Biology of Plagues: Evidence from Historical Populations.* Cambridge: Cambridge University Press, 2001.

Sheets, Bob, and Jack Williams. *Hurricane Watch: Forecasting the Deadliest Storms on Earth.* New York: Vintage, 2001.

Smolin, Lee. *The Trouble with Physics: The Rise of String Theory, the Fall of a Science, and What Comes Next.* Boston: Houghton Mifflin, 2006.

Sunstein, Cass R. *Laws of Fear: Beyond the Precautionary Principle.* Cambridge: Cambridge University Press, 2005.

Winchester, Simon. *A Crack in the Edge of the World.* New York: Harper Perennial, 2005.

Zebrowski, Ernest. *Perils of a Restless Planet: Scientific Perspectives on Natural Disasters.* Cambridge: Cambridge University Press, 1997.

Zebrowski, Ernest, and Judith A. Howard. *Category 5: The Story of Camille, Lessons Unlearned from America's Most Violent Hurricane.* Ann Arbor, Mich.: University of Michigan Press, 2005.

INDEX